This book is for my Family

To the memory of

John Dalton

English Chemist 1766 – 1844

My Chemistry Students at QHIS

Copyright © 2009 Tom Law

2nd Edition Jan 2016

ISBN   9780994315755

Published in Australia by:

Longership Publishing Australia

3896 AUSTRALIA

ABN 73446736413

email: longership@email.com

Copyright © Tom Law   2009

Second Edition Jan 2016

Cover design Tom Law

Law, Tom

Elementum Carbone   ISBN: 9780994315755

pp 264

TOM LAW has been a science and computer teacher for most of his professional life. Born in Glasgow, he migrated to Australia at age sixteen with his parents and two brothers. He has worked in schools in Melbourne, rural Victoria, Indonesia and China. He has always maintained a strong

photo: Zhang Yumin

interest in the environment and issues surrounding the nuclear industry and climate change. He is never afraid to speak out against injustice and identify those regimes pursuing the undemocratic road. This is his second book. Tom is married with two children and currently lives in Changzhou, Jiangsu Province, China. He has two adult sons from his first marriage.

ACKNOWLEDGEMENTS

It seems unfair to mention just a few names here when in fact so many have guided me and provided both material and critical commentary assistance. So I say to those friends not mentioned 'please forgive me'. The whole is the sum of innumerable parts and it is often difficult, indeed impossible, to source the genius who first described an idea or made some unique and original quote pertinent to the subject at hand. Such is the nature of the collective science and knowledge of humankind.

Miss Yang Weiwei, Changzhou China for assistance with source materials

Mr Shen Feng, Changzhou China also for assistance with source materials

3

Dr Robert Wm Johnston, wwwjohnstonarchives.net for commentary on the problems with the science of climate change

Department of Biological Sciences, University of Virginia, undergraduate notes

Department of Geology and Earth Sciences, University of Virginia, undergraduate notes

Image Science & Analysis Laboratory, NASA Johnson Space Center, http://eol.jsc.nasa.gov

Nanjing Institute of Geology and Mineral Resources for fossil specimens, Hu Nan, China

Eric Roston, author of The Carbon Age, Walker & Company, whose work I was ignorant of when I first commenced to write this book but later provided insight to another perspective.

Bogor Botanical Garden, Bogor, Indonesia.

A special thanks to the Principal, Staff and Students of Qian Huang International School, Wujing, Changzhou, JS, China.

Zhang Yumin for the photograph of yt.

Many more references have been cited in the Bibliography at the end of the book.

Elementum
Carbone

John Dalton 1766 - 1844
(from an engraving by Worthington)

Tom Law

Longership Publishing Australia

Contents

Introduction

In my book "Nuclear Islam and Other Stories" - Sid Harta Publishers Australia 2007 – on page 72 I depict the availability and demand of oil and predict that it will all be gone by the year 2073. Flicking the page, there is a graph showing the combined resource of wood, coal and uranium with two curves: one for demand and one for availability. Availability peaks at about 2035 and the two curves cross at about the year 2075. These statistics presume of course that the world economies will continue on their current high road of madness and extravagance. Frankly, I don't see why they will not continue in this vein.

What is the reason for the world and its peoples to be phutted[*] by around 2075? I have thought hard about this and can come only to a simple but plausible explanation. 'The whole world is controlled by

* see Coda of same book for an explanation of "Moment of Phut"

accountants and economists.' Further, I have personally not met an accountant or economist that has an in depth understanding of his own science and has no knowledge at all of mathematics or other technological sciences. They know that 1 + 1 = 3 (a must to graduate in these disciplines) and that 1 – 1 = catastrophe! Colours are reduced to red and black with the number ' zero' instilling fear and insecurity, engendering palpitations and even heart failure! The 'black figures' in the books together with the shareholders are the driving forces behind all decision making across the globe. Executive boards and managing directors of global corporations must ultimately surrender to the advice of these not so gentlemanly gentlemen. 'We have run computer models and crossed referenced with all the market indicators and fiscal probabilities, recognising human resource undertow together with current currency trends and trigonometric cyclical forecasts. Taking a parsimonious selection of 0.03 in the gamma determinant of the superimposed exponential bias, our recommendation at this time is not to make any hasty adjustment to our presently agreed position."

*So what the 'Phut' does all this gobbledygook mean? The real problem is that neither any of the board nor the proponents (our so-called chief economic and accounting advisors) understand any of this either. It is merely the secret language of the board-room akin to Latin for doctors, attorneys, pharmacists and Catholic priests purely as a tool to befuddle the minds of the masses. There may be a single egg-head hidden away somewhere in the basement below the car park in a bunker that produces the mathematics etc but the real interpretation of the whole message is simply: **business as usual!***

The true powerbrokers of the world are only interested in profit and to maintain the status quo they will wield their power in the corporate scene, the media and the parliaments of all the democracies. The environment, finite resources and the detrimental effects of industrialisation are always secondary at most and placed on the back burner generally. They will take us right to the wall never looking back... they lack a social conscience and see profit and balancing the books as the only criteria in a tunnel view of the present and the future of humanity and planet Earth. They will lead us to the final abyss from which there will be no escape, rendering death to our civilisations.

The repression and starvation of the poor, the spilling of poisons from the vials, the seas and oceans turning red, the sun and moon barely seen through the smog of a polluted world. These images from Revelation are familiar enough to most but it does not have to be this way. It would be foolish to turn our heads and say "but what can I do?" And that is the very essence of this writing: 'that each and every individual has a responsibility and must act and behave in the coming time to turn around and away from the downhill path of choking pollution and destruction of our beautiful planet.'

Only by a collective response by all the nations and the peoples of the world can the process of rapid degradation be halted and reversed to heal and beautify again, regaining the splendour of a planet unique among the cosmos. Crystal streams, blue skies, forested mountains and bountiful oceans can all be achieved and necessary if we are to avoid 'the oxygen dive'. It is the kingdom and domain of humanity and humans reign and control all aspects of the land, the sea and the air we breathe. Too much has been squandered on armaments that can never be used. Too much has been squandered for personal gain without a view of the larger picture. Long term planning for our civilisation is not on the agenda. Five year plans in this instance are inconsequential as far as preserving the environment. We need thirty year plans with a five year overlap superimposed on a hundred year plan.

Education together with the behaviour and values of the individual are paramount. This is in place to some extent in our schools. But it is personal empowerment that is the key. Governments around the world must recognise this and provide funding and scheme development to aid communities and individuals to better their life pattern, clean their yards, put in place alternative energy generation and create scientific organisations to study the problems to render solutions to create a secure and sustaining future for our children and their children. Let us not cry out "woe is us… the curse of the apocalypse is upon us!" We are alive and with our combined intelligence we can avert doomsday. It is our duty to stand up and fight the doomsday prophesiers and be appalled by actions that will lead to the fulfilling of such prophecy. We must act now. We must act together. Through careful analysis, trial and prototyping, looking to the long term and final application of proven strategy we may inch forward to a new world. Let us not ignore our experience of the past or throw away our history. Instead let us learn from it but keep it recorded for all time that future generations might look back and be blessed and thankful for our vision of what was and what can be with a will and determination. Let this be the hallmark of the beginnings of the twenty first century. The time of the pleasure seekers and hedonist is passed. The world is full of work to be done, work that must be defined and enacted in new and revolutionary ways. Humankind likes a challenge. This then is now the greatest challenge in all time.

Changzhou, China May 2009.

Notes to this Edition: *I have changed almost nothing in this printed edition from the original ebook. I have made only perfunctory changes with minor editing here and there where I desired more explanation or deemed the language needed improvement. However, emphasis must be placed on the critical state of our atmosphere and oceans even now in 2015. Despite our knowledge and obvious predictions, the expansion of mining and consequential usage of coal alone is to increase six fold over just the next five years to 2020! Oil and gas will follow a similar trend consuming oxygen and creating an unbearable load of carbon in both the air and in the world's oceans. Concrete and steel will also continue to expand in production, further depleting atmospheric oxygen. Will we reach the dreaded moment of 'flip' where a runaway cycle takes hold reducing all the species of living things upon the Earth, It is the higher forms of life such as the mammals that will be struck the hardest. From a distant and unemotional viewpoint, the planet has seen this sudden demise of its life forms before in a catastrophic and sudden manner. The planet will go on. But does humankind wish to bring this upon itself? Can the big and powerful take heed or continue to drift towards the precipice? We shall see soon enough!*

By the way, if you find Chapter Zero is a bit difficult to understand in parts and get bogged down, there is an easy solution… just skip it!

East Gippsland, Australia Dec. 2015

Conquest, War, Famine and Death are illustrated in the famous woodcut by Albrecht Durer

Zero   **Life Before Carbon**

In the beginning God created the heavens...

Darkness was upon the face of the deep...

And God said, "Let there be light" and there was light.

And God saw that the light was good.

Of course there was no life before carbon. How could there be? Before carbon there were no elements or at least not much more than hydrogen and helium. And before hydrogen and helium? There is the 'big bang theory' which cosmologists religiously adhere to. That is, that matter was suddenly created from nothing in a mighty explosion whose expansion is still evident today (expanding universe theory) and the residual background microwave radiation or universe temperature from that initial explosion can be detected (today, the background radiation has been measured at 2.73° Kelvin.) But these three things, hydrogen and helium; expanding universe; and residual background radiation, each have problematic alternative explanations which may or may not disprove the big bang per se. However, having said this I

am not truly taken with the 'steady state theory' that says the universe has always existed. One of the irritating things is how promoters of the big bang refer to 'the first 3 minutes' or 'the first nanosecond' as if time is somehow estimable in the process. I will return to this later.

Problem of the First Cause: *(also refer to Appendices E and F at end of book)*

In religious writings and doctrine, the first cause is attributed to the hand and mind of God. I am not here to argue against one's personal faith. However a question put forward by cosmologists might be 'Where did God come from?' Of course cosmologists fall into the trap that the universe, and outside the universe, is somehow governed by time. But I postulate later that time is a concept of relativity and gravitational field (in turn, matter) and not directly related to experience and to the lifespan of organic living entities. Time is inseparably tied to matter as space itself is also tied to matter. No matter, no space and no time!

Our laws of physics, when describing fields such as electromagnetic, electrostatic and gravitational have invented the mathematical concept of infinity, a useful idea. Space, with reference to field laws is often considered to be infinite. But I question that here. I describe space as being an entity that is dependent on the existence of matter. In this sense space is a product of gravitational field hence it, itself, is the propagation and extension of matter. I also theorise that it is two dimensional at great distances (with a bump at the middle) that curves back on itself akin to the surface of a sphere, whose mean radius is dependent on the quantity of matter. There is no inside or outside. There is no boundary to something else external. It is complete in itself.

Black Hole Bubbles

The big bang theory follows from "as the universe appears to be expanding, it must have started from a single point". But this argument is not strong although many would at first take it to be self evident. A balloon does not start from a point but we may cause it to expand. There might be many reasons for a real or apparent expansion. Can something as simple as photon pressure account for expansion? Also, as photons accelerate around stars or galaxies i.e when passing through stronger gravitational fields, do their wavelengths not elongate? Thus the further a photon travels it would experience a red-shift due to its passage through many gravitational fields. A black hole is formed by a concentration of matter to such an

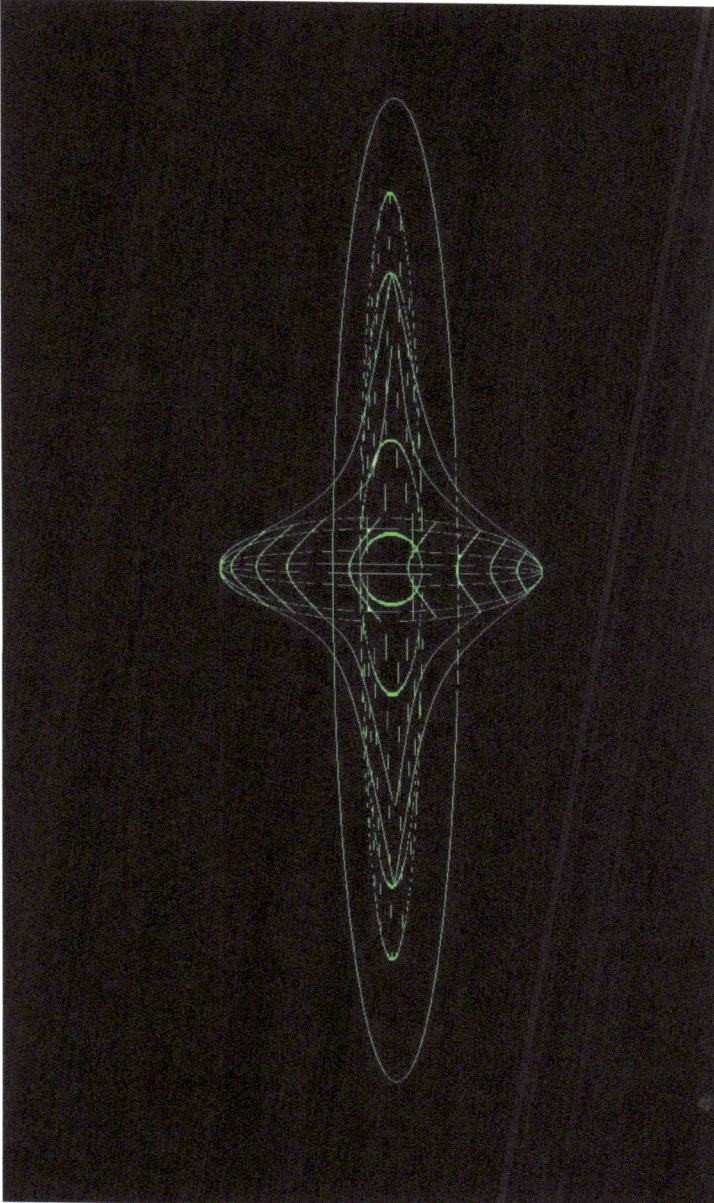

The universe is predominantly two dimensional with a bump in the middle!

extent that matter collapses under its own gravitational force to the extent that light cannot escape. Hence the term black hole. But where does this matter go to? One suggestion is that it creates a new space in a separate dimension, the only connecting point being at the centre of the black hole. Thus the original space is contracting whilst the new space is expanding as the black hole gathers material:

The original space is contracting whilst the new space is expanding. Does this imply that the universe La Grande contains many

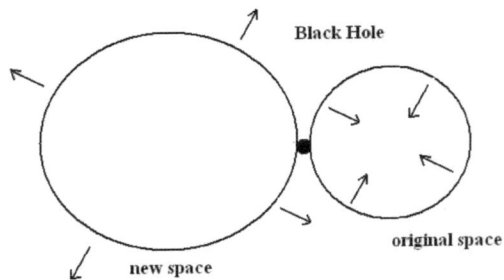

unconnected spaces other than at the dimensionless singularity at the centre of each black hole? If this process were to continue then the current space we know as our universe will eventually disappear!

But it might be that our universe is also being 'fed' with matter via black holes from other spaces or another space. The net effect may be a balance. I still haven't got to the discussion on the first cause. I must mention a few things on the idea of time.

Time Quantised but Elastic

I postulate here that time is quantised; that is in any gravitational field due to matter there is a limit to a change in state of relative orientation of objects. I am regarding the passage of time here as akin to the flipping of cards or frames where each frame represents the relative position of particles at the quantum or ultimate level. There can be no intermediate positions as length is also quantised. For example, suppose a universe consists of nine available positions and we have just two distinguishable particles labelled black and red. With the black in the top left corner, we immediately identify eight states, depending where we find the red particle.

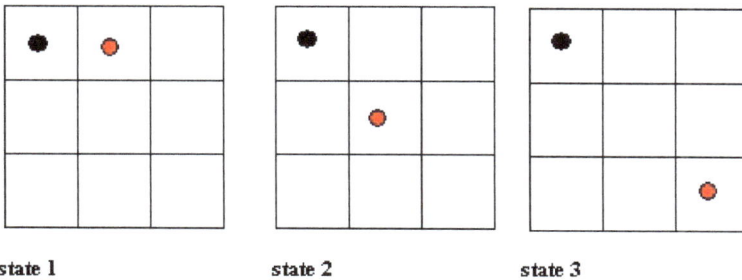

state 1   state 2   state 3

Now in this simplistic universe we assume that the two particles may move about relative to each other and we find that we have a total of 72 possible states. What is the meaning of time here? The system may pass through successive states which we may assume are random until such time the system is once again in **state 1**. it will cycle again taking more or fewer steps than before until it reaches the initial **state 1**. For each cycle we might say that the states traversed until the initial state is reached are countable. Now lets introduce a second 'connected' space but with no passage of particles between them. This second space has only one particle and two positions possible with the distinction that it may flip or change state more frequently relatively to our first space (we might say it has a stronger gravitational field) We see in the figure that space 2 (Γ_2) has cycled through three states whilst space 1 (Γ_1) has not yet changed state. Space 1 changes state as *space 2 changes for the fourth time. Let us suppose that these two spaces are not really separate but merely subspaces of the one total space Γ , but with some different field influence such as gravity.*

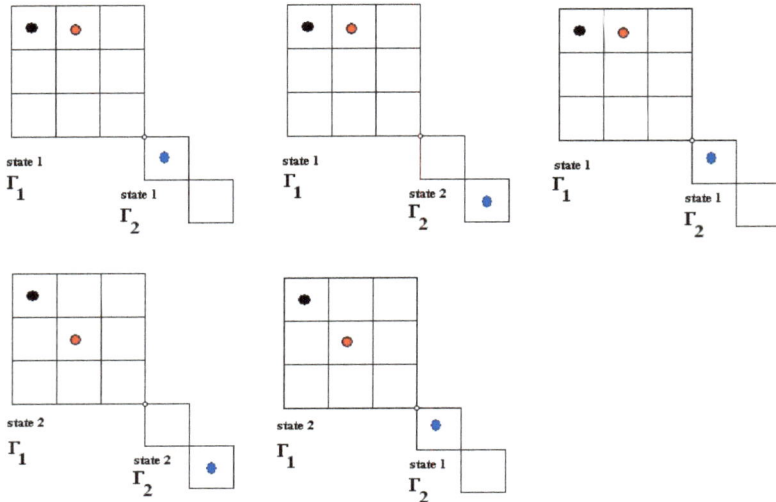

So what am I trying to convey here? Firstly that time is not real in the sense that we imagine it to be in everyday experience. It is a trick of physics due to the complexity of the universe we live in. The reality is that time is a sequence of physical states and that in the simplest of models, states return to the starting point after a cycle which may be subject to both random and descriptive effects. In the real universe we see an ever complex pattern of states which we term evolution. Mathematically speaking it is difficult to see that the universe can ever roll backwards to an earlier state that it has already seen. However this may not be so if we imagine the idea of contraction and destruction via black holes to simpler primordial states. It would appear then that as time does not flow as imagined in life experience,

we can never return to an earlier state or move very far forward to a later state as in science fiction. However, Special Relativity does allow us to move forward to later states, but at present there seems to be no physical or theoretical means to return to an earlier state in a complex universe as was demonstrated in the previous simplistic model. But the very essence of the simplistic model does not rule out the possibility of moving to an earlier state. However there must be consequences that may be irreconcilable and thus again leading us to conclude that it is not possible.

So when I observe a quasar at the theoretical limit of space I am looking at the ghostly afterglow of an object that was there billions of years before. It would be like having a room full of mirrors some of which I could look into and perceive my own birth and childhood due to the light rays creeping across the space in the room at a crawling pace. So my one galaxy universe seemed just as plausible as a many galaxy universe due to the very nature of its size and finite dimension coupled with its curvature.

Where did carbon come from?

The universe is made up mainly of atoms and ions of the elements hydrogen and helium together with electromagnetic radiation in the form of photons. Some of these we can detect with our eyes, however many cannot be seen unless we use some special detector. In the outer reaches of galaxies we see the accretion or appearance of matter which is slowly drawn in to the galaxy. These vast clouds of hydrogen, stretching for hundreds of thousands of lightyear (ly), coalesce into stars under the attractive force of gravity. How and why this matter is formed in empty space is not fully understood, but must come from photons that have travelled across the universe and are somehow captured or slowed to pass through a complex series of mutations to form basic building blocks of matter. We know that the centre of the Milky Way has a considerably greater density of black holes than in the spiral arms. The shape of our galaxy suggests a repeating pattern from solar systems... a central large mass with matter orbiting in a plane rather than in all directions. This is how I view the whole universe but again with no outside or boundary. Difficult to imagine other than by stating that whichever direction one tries to follow, eventually one would arrive at the starting point assuming the passage was instantaneous. The universe is a big place and when we look out into space beyond the galaxy we can only perceive history of eons past. I once stated that perhaps our galaxy is the only galaxy in existence and that all else is an illusion!

One Galaxy in a Two Dimensional Universe

Godstar

This might give you some amusement but I pose it here as a serious hypothesis. The value is an understanding of the nature of the universe, so that alone makes it an important tool. One night whilst lying in my bed in Kemanggisan, a suburb of Jakarta, I noticed the reflected small red diode light from the light switch. There seemed to be an excess of mirrors due to glass over pictures hanging on the walls and the internally mirrored wardrobe doors lying open. I was amazed at the multitude of red lights I could see from reflections of reflections of reflections all emanating from the one single source. I got to thinking of Einstein's General Theory, as one tends in such situations, and wondered if the universe was just one big con due to gravitational perturbations, refractions and reflections of light. All the galaxies we see, I pondered, might just be our own at various stages of its evolution. Further, in my possibly inebriated stupor but wanting in sleep, I developed a theory of a massive central black hole which I named Godstar around which our galaxy revolves, expanding to its aphelion (for want of a better term) and contracting to its perihelion.

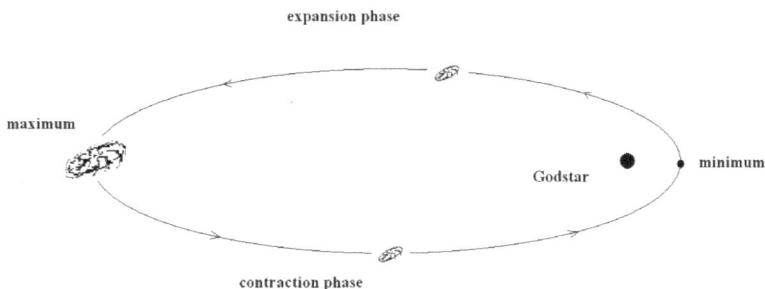

I assumed that Godstar must be several orders of magnitude bigger than our galaxy (referring to total mass). Everything else we see is

just an illusionary ghost and shadow of our past selves, all being purely gravitational reflections and refractions. If I am correct, our nearest neighbour, Andromeda will just suddenly disappear one day to the astonishment of all! I soon after fell asleep and to this day can recall no more details. But the study of a single atom universe or a single galaxy universe can be a useful tool that helps in our understanding of the nature of the beast i.e the real universe!

A Repeated Pattern.

*Saturn has beautiful rings around it as we all know. The dimensions of the solar system are specified in terms of the mean distance from Earth to the Sun, called the **astronomical unit (AU)**. One AU is 150 million km. The most distant known body orbiting the Sun is the dwarf planet **Eris**, whose discovery was reported in July 2005. Eris is currently about 97 AU from the Sun. Another planet-like object in the outer solar system is named Sedna and is currently at 90 AU but will reach about 900 AU at the farthest point in its orbit thousands of years from now. Comets known as long-period comets, however, achieve the greatest distance from the Sun; they have highly eccentric **orbits** ranging out to 50,000 AU.*

The planets Jupiter, Uranus, Neptune and Pluto all have rings orbiting around them. Saturn's rings are by far the largest and most spectacular. With a thickness of about a kilometre, they span up to 282,000 km (this is equivalent to about three quarters of the distance between the Earth and the moon.)

Now the observation is that this "ring thing" seems to be repeated throughout the visible universe. Whereas we do see globular galaxies, the main seem to be flat spiral with a bump in the middle. The solar system has virtually all its planets and other material rotating in the same plane extending out to the Kuiper Belt and Oort layer. (Pluto is a minor exception with an inclination of some 17^0 to the ecliptic plane) Our own galaxy consists of about 200 billion stars, with our own Sun being a fairly typical specimen. It is a fairly large spiral galaxy and it has three main components: a disk, in which the solar system resides, a central bulge at the core, and a surrounding halo.

The disk of the Milky Way has four spiral arms and it is approximately 300 pc thick and 30 kpc in diameter. It is made up predominantly of Population I stars which tend to be blue and are reasonably young,*

spanning an age range between a million and ten billion years.

The bulge, at the centre of the galaxy, is a flattened spheroid of approximate dimension 1 kpc by 6 kpc. This is a high density region where Population II stars are prominent i.e stars which tend toward red and are much older, about 10 billion years. There is evidence for the existence of a massive black hole at the centre and possibly other black holes within the bulge.

The halo, which is a diffuse spherical region, surrounds the disk. It has a low density of old stars mainly in globular clusters (these consist of between 10,000 - 1,000,000 stars). The halo is believed to be composed mainly of dark matter which probably extends well beyond the edge of the disk.

* parsec: cosmological unit of distance. See next section:

The Astronomical Unit, the Light Year and the Parsec:

The average distance between the Earth and the Sun is a natural choice for a unit of distance and is called an Astronomical Unit (AU).

$$1\ AU = 1.5\ x\ 10^{11}\ m$$

The light year (ly) is another often used unit of distance

$$1\ ly = 9.5\ x\ 10^{15}\ m = 6.3\ x\ 10^{4}\ AU$$

Inter-galactic distances are often expressed in the larger unit of distance called the parsec (pc). This is defined as the distance at which one AU would subtend an angle of one second of arc:

$$1\ pc = 3.1\ x\ 10^{16}\ m\ = 3.3\ ly\ =\ 206000\ AU$$

A nearby star's apparent movement against the background of more distant stars as the Earth revolves around the Sun is referred to as parallax.

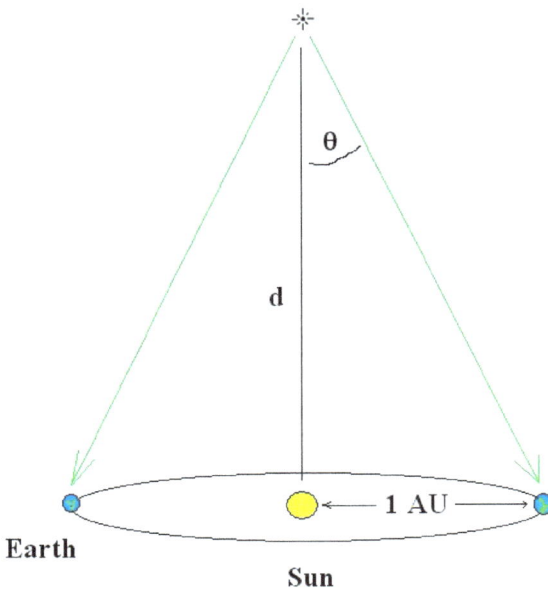

The parallax can be used to measure the distance to the few stars which are close enough to the Sun to show a measurable parallax. The distance to the star is inversely proportional to the parallax. The distance to the star in parsecs is given by:

$$d = 1/p$$

where d is in parsec and p in seconds of arc (1/3600 of a degree)

The nearest star Proxima Centauri has a parallax of 0.762 arcsec, giving a distance of 1.31 parsec.

Coming back to my main point then is that galaxies also display a predominantly flat disk shape with a bump in the middle (but with exceptions as observed). This disk is now believed to extend far out into space in the form of dark matter.

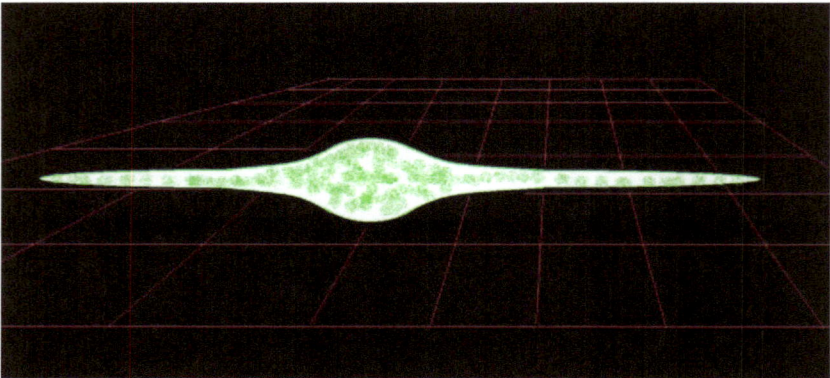

So my hypothesis is: "Most galaxies cluster in some disk shape with a bulge at the centre". Why can we not see this by observation? Well there are two main factors: (i) the extent of the universe i.e its size and (ii) the warping, refraction and reflection of light from distant objects due to gravitational fields.

What is space anyhow? I tend to think that space is a function of both matter and field.

$$\Psi \sim X, \Phi$$

But field is also a property of matter and consists in the main of gravitational and electro-magnetic

$$\Phi \sim \Gamma, E$$

Therefore I feel that space is defined by field. No field, no space! More exacting, no matter, no space! From this point of view, space is more or less two dimensional with just a blip in the middle. An important blip none-the-less! Admittedly we fly in aeroplanes and climb up and down stairs in our daily lives. But most of our ventures are in two dimensions over the surface of the Earth.

So what happens at the edge you ask? Well there is no edge in effect. Electromagnetic field is not as extensive as gravitational field. Even flatness is deceiving... space must curl back on itself to form a closed continuum just like the surface of a ball. We cannot see outside because there is no outside. We are never aware of an edge because whichever direction we take we eventually arrive back at our start point. Conceptually difficult? Yes.

Scientists have calculated the existence of dark matter at the edge of the galaxy and throughout space generally. Why? Well the current visible mass just does not add up to make the whole fabric hang together. Some estimates are that 90% of the universe is this dark matter. We know that at very low temperatures, it is

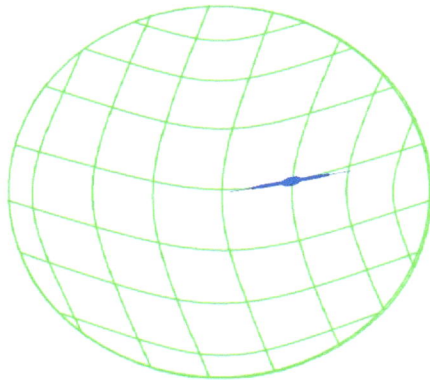

possible to slow photons to a standstill. Why can't dark matter be stationary (relativistically) photons? Can photons evolve into matter by quantum processes in intergalactic space?

Science fiction writers are always looking for the additional higher dimensions but because of the curvature of space it seems that we have fewer dimensions. Time is often considered as the "fourth dimension". This is convenient in our every day experience as we see ourselves getting older and change going on around us. Does it have such significance on the quantum level? At that level we seem to be more concerned with "states". See earlier for an explanation.

Life and the First Crawl

A study of nuclear interactions, particularly the synthesis of the elements in stars due to nuclear fusion and as a result of supernova, we see that over eons, the whole gamut of elements and their isotopes are slowly formed. But carbon is central to those structures for living things with the presence of oxygen and nitrogen. How these elements are synthesised in stars is a series of complex nuclear reactions more detail of which is in Appendix B. The structure of a planet like the Earth with iron, uranium and so many useful heavy elements have evolved over many cycles of stars reaching there life's end then exploding into a super nova, flinging stellar material far and wide to coalesce again into a new star perhaps with planets. To create life in its simplest form we need carbon, hydrogen, oxygen and some phosphorous and nitrogen atoms. The simplest of proteins are formed from various alignments and combinations of amino acids.

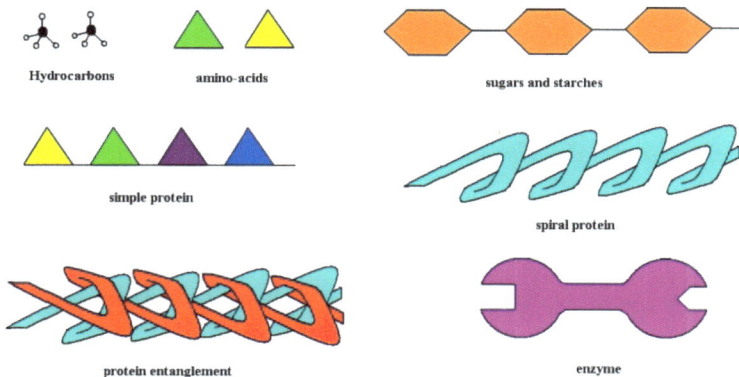

Hydrocarbons   amino-acids   sugars and starches

simple protein

spiral protein

protein entanglement   enzyme

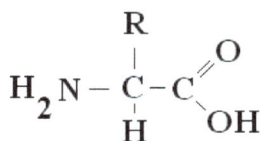

where R is hydrocarbon (alkyl) group

basic structure of an amino-acid

It is argued that even the age of the Earth, 450 billion years, is statistically not even close enough for the evolution of these basic building blocks of life. The age of the Universe is currently estimated at 13.5 Gyr (a giga year equals one billion years). Life on Earth is estimated to have commenced in its most primitive form some 3.5 Bya (billion years ago). This provides a time scale of 1.5 billion years to create the first prototype of life. Possibly, if some of the essential building blocks were seeded on the Earth say from meteorites the time scale is sufficient. Meteorites have been found to contain hundreds of amino-acids; this is far in excess of the essential twenty found in living things on this planet. Only in the large nebulae (clouds of gas and *particles) in the Universe does there exist sufficient factors (such as age and sheer volume) where chance collisions eventually sequence the necessary amino acids to form the simplest proto-life forms. The question remains, did the first living thing form in these interstellar clouds or was this possible only on the planet? Carbon and carbon monoxide are essential elements to commence the synthesis of those compounds required for life forms to be created. Hydrocarbons, amino-acids and simple proteins are required next. We know that some proteins form spiral arrangements due to weaker bonds formed between their component atoms, behaving like invisible threads holding the spiral or spring structure together. Protein chains may become entwined. Proteins may then form the simplest of DNA (deoxyribonucleic acid), the double helix molecule that can unravel and form two new DNA molecules. This is the basis of replication, essential for the continuity of living organisms.*

We see here that element 15, phosphorous, is necessary along with sugar molecules to form a nucleic acid. The double helix structure must be triggered to unravel and form two new structures. This is the basis of reproduction which is a key feature in distinguishing something which is alive compared to something which is inanimate.

The human genome (i.e DNA) contains 3 billion base pairs and up to 35000 genes, give or take a few. The retrovirus Rous Sarcoma contains as few as 3500 base pairs and just four genes. Its size is around 80 nm in length. So we are looking for a 'Microplasma Nebulum' around say 15 nm and containing 600 base pairs and just a pair of genes. Will we ever find such a primitive life form? If we do, it could be argued that all living things with all their complexity

Molecules that Replicate, the Nucleic Acids

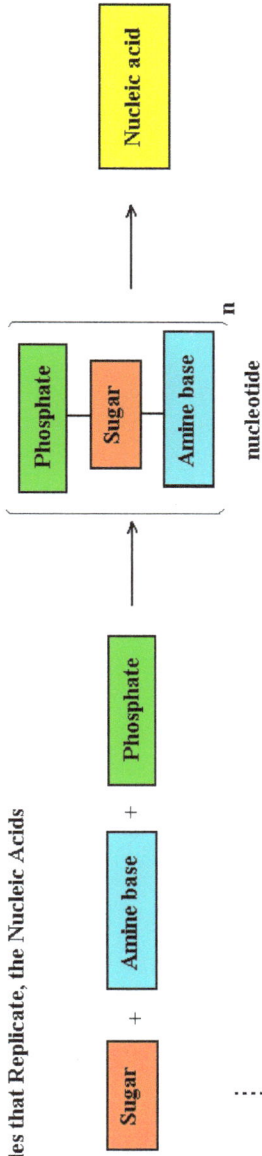

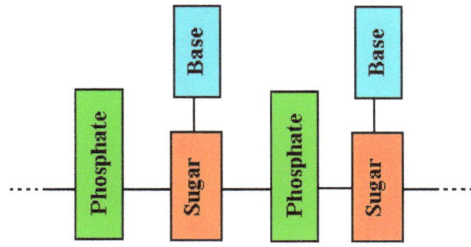

Sugar + Amine base + Phosphate → [Phosphate — Sugar — Amine base]ₙ → Nucleic acid

nucleotide

doublehelix structure of DNA

Phosphate — Sugar — Base
Phosphate — Sugar — Base

Structure of DNA

32

DNA Replication

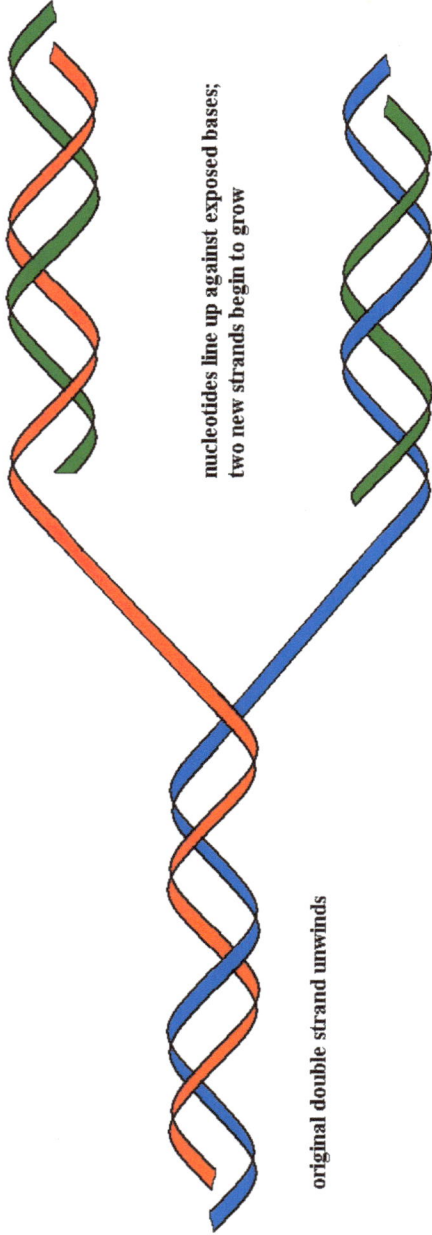

nucleotides line up against exposed bases;
two new strands begin to grow

original double strand unwinds

Crick described the process by an analogy of a hand and a glove. The hand and glove separate, a new hand forms inside the glove and a new glove forms around the hand! two identical copies of the original now exist.

stemmed from this first replicating thing on the boundary of chemical molecule and living creature; our genetic ancestor. Of course to be a truly living thing it needs some containment barrier or cell wall. Also it must consume energy and organic material which we call nutrients; but replication of itself is foremost and the gradual development of proteins and enzymes to perform symbiotic functions within the boundary would appear over many millions of years as the 'protobiotic' evolved.

So the 'Microplasma Nebulum' were swept along by meteorites, comets and similar objects eventually to be deposited on planets. If conditions are favourable then the planet is seeded and evolution of living things begins.

Looking at the trillions of star systems in our galaxy alone it is most unlikely that life is unique to the Earth. We will know soon enough. Here is a quote from an article:

"So our understanding of the universe has undergone a drastic overhaul. The simplistic and placid world of Ptolemy and the ancient Egyptians, with a tranquil Earth at the centre orbited by the sun and adorned by the presence of other planets. Copernicus overturned this idyllic vision which had lasted millennia and suddenly the Earth has been relegated to a child of the sun gadding up and down the universe. Today we have made a further leap towards a neurotic cosmos, towards a universe which tends towards an infinite dispersion, sudden fruit of an enormous catastrophe Beyond the Milky Way, once mistaken for the immutable expression of an eternal order, there are apocalyptic dramas being played out repeatedly: stellar explosions, collisions

of stars and comets, encounters between galaxies. The sun, moon and stars, faithful companions of merchants and navigators over millennia, are time bombs, patiently waiting their inevitable end, finishing as they started, in catastrophe. Fear of the apocalypse has given way to mathematics and the cold, rational certainty of this inevitable end. Divine providence has packed its bags, hope is dead and buried. The galaxies are born and die like prehistoric animals, like cosmic dinosaurs. The real universe, a far cry from only order and peace, boils, is agitated and flaps in spasms of geneses and convulsions of agony. The depth of the blue coloured sky, where we had transferred Olympus, is today the theatre of nova, supernova and of their explosions. Elementary particles, quarks, atoms, ions and molecules with no fixed abode alternate with storms of deathly rays. Condensation of energy into matter, and vice versa the annihilation of the same material are the order of the day. Flashes of mortal light dance across the dense clouds of dark matter, of gasses and of cosmic dust which obscure our view. These clouds will sooner or later collapse giving life to new stars and planets.

This frightening picture, conceived one piece at a time after the progress of astro-physics and physics. Spectroscopy, already amply used in the study of the structure of materials, and radio-astronomy, used to scrutinise space, has demonstrated that empty interstellar space hardly exists. Space contains a confusion of complex organic and inorganic molecules. Surprising results have been reached thanks to recent measurements made outside the Earth's atmosphere, a fortunate filter of space radiation. We know that at thousands of light years from us, there are many of the composites necessary for the synthesis of macromolecules like proteins and peptides, products which play a highly relevant role in the appearance of life. We also know that these "biological bricks" existed in space before our solar system and our planet. Therefore, life most probably came to Earth from far away shores, or developed on Earth and other hospitable planets, parting from the same building bricks. The discussion on how and where life arose is still open. On this question, which has an antique flavour, we later propose a new reply, even if we remain the ashes and dust of stars."

The Garden of Eden of course was not the land itself, but for our planet the vast expanse of oceans. Geological evidence points to this fact that the earliest forms of living things followed by an explosion of variety and diversity occurred mainly in sea water. At the boundaries of seas and dry land and in shallow waters, the land plants took hold. The part played by insects cannot be underemphasised. Again this process of assistance in cross-fertilisation occurred in shallow seas but the development of winged insects favoured the blooming and spread of grasses, shrubs and eventually forests of giant conifers across a young and fertile landscape. With the expansion and contraction of polar ice together with volcanic action, the flipping of the magnetic field and many other major physical influences plants insects and animals filled the atmosphere with odours, seeds and noise. As these mighty forests grew they deposited carbon layers in rock strata. The shell fish deposited carbonates of calcium and the simplest of ocean creatures filtered down to be squished into petroleum oil. Carbon spread over the surface of the planet and under the sea in the form of carbonates, oil and coal and lay undisturbed for four hundred million years. Only the natural interaction of volcanic lava and coal seams caused the ignition of this fossilised material but the majority lay quiet, awaiting the emergence of man and the industrial age.

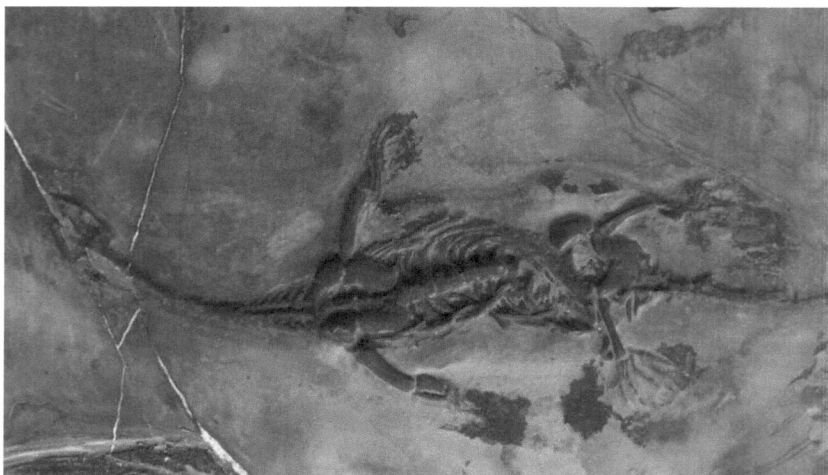

One   *Evolution of the Earth's Atmosphere*

You might like to study geological time and how the various animals, plants and marine organisms evolved. We have much evidence to support our estimates of the age of the solar system and the Earth in particular. Life has slowly developed over the past 570 million years from the most primitive of single celled organisms to the complexity of mammals to which class humankind belongs. This time period pales to insignificance compared to the four and a half billion years which is said to be the age of the Earth itself. So there existed a long time when there were no living things crawling about and making a nuisance of themselves.

The atmosphere of the Earth is generally believed to have its origin in relatively volatile compounds found in the solids from which the planet was formed. Such compounds included nitrides, ammonia, water, carbides, and hydrogen compounds of carbon such as methane. Many of these compounds can form minerals which are fairly stable at low temperatures. The high temperatures developed during the later stages of accretion as well as subsequent heating produced by the decay of radioactive elements released these gases to the surface of the planet. Down to the present time, large amounts of CO_2, water vapour, N_2, HCl, SO_2 and H_2S are regularly emitted from volcanoes. The more reactive of these gases were removed from the atmosphere by chemical reactions with surface rocks or dissolution in the ocean, leaving an atmosphere enriched in its present major components with the exception of oxygen. Hydrogen, being so light, would tend to escape into space, causing the atmosphere to gradually become less reducing chemically. The lighter noble gases helium, neon and argon, which are among the ten most abundant elements in the universe probably underwent fusion in the early formation of the planet. Whatever the cause, we find these elements in extremely low concentrations compared to the universe as a whole (they are depleted on the Earth by factors of 10^{-7} to 10^{-11}).

The overall oxidation state of the Earth's mantle is not consistent with what one would expect from reactions with highly reduced volatiles, and there is no evidence to suggest that the composition of the mantle has not remained the same.

The primitive atmosphere is thought to have consisted mainly of water and carbon dioxide together with smaller amounts of other gases.

If water vapour was a major component of gas formation of the young Earth, a proportion would have condensed into rain. However the significant concentrations of water vapour and carbon dioxide

Formula	Name
H_2O	water
CO_2	carbon dioxide
N_2	nitrogen
H_2	hydrogen
H_2S	hydrogen sulphide
SO_2	sulphur dioxide
CO	carbon monoxide
CH_4	methane
NH_3	ammonia
HCl	hydrogen chloride
HF	hydrogen fluoride
NO	nitric oxide
NO_2	nitrogen dioxide

in the atmosphere would have led to a runaway greenhouse effect, resulting in temperatures as high as 400°C in the early stages. To *solve this problem is difficult. It is likely that the Earth–Moon interaction has played a significant role, particularly with regard to the early Earth's rotational speed. The theory poses more questions the answers to which may be modelled. However a fully comprehensive model must be swallowed with a fair amount of parsimony.*

Free atmospheric oxygen was merely a trace component of the youthful planet. Oxygen is much happier when combined with other elements as oxides in rocks and minerals. Photochemical decomposition of gaseous oxides in the upper atmosphere is the major source of O_2 in the pre-biotic state. On the Earth, the major inorganic source of O_2 is the photolysis of water vapour with most of the resulting hydrogen escaping into space, allowing the O_2 concentration to build up.

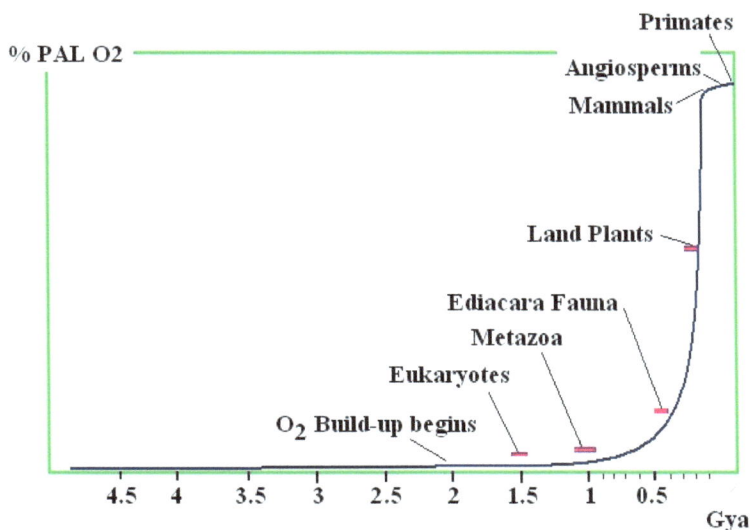

At an estimated 2 x 10^{11} g of O$_2$ per year, this amounts to less than 3% of the present oxygen abundance. The major source of atmospheric oxygen then was due to photosynthesis carried out by green plants and certain bacteria:

$$6H_2O + 6CO_2 \rightarrow C_6H_{12}O_6 + 6O_2 \text{ - ENERGY}$$

A view of the build up of atmospheric oxygen concentration since the beginning of the sedimentary record (4 x 10^3 Mya)* can be estimated by making use of the fact that the carbon in the product of the above reaction has a slightly lower C^{13} isotope content than does carbon of inorganic origin. Isotopic analysis of carbon-containing sediments thus provides a measure of the amounts of photosynthetic O$_2$ produced at various times in the past.

[* Mya : million years ago Gya: thousand million years ago]

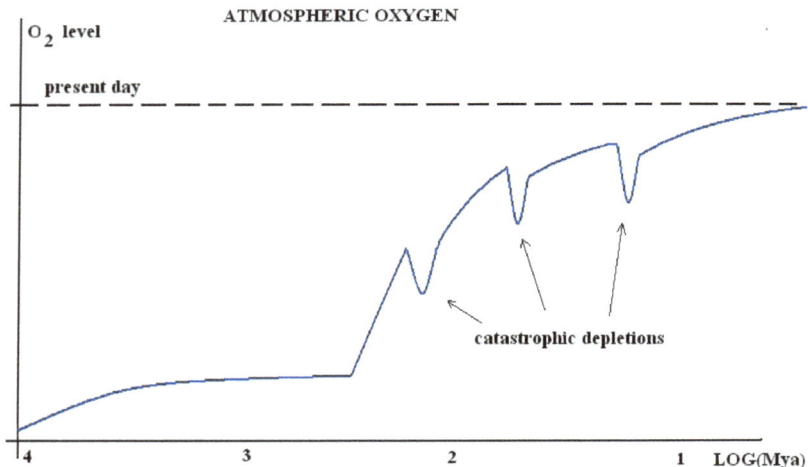

ATMOSPHERIC OXYGEN

O$_2$ level

present day

catastrophic depletions

4   3   2   1   LOG(Mya)

41

Carbon dioxide has probably always been present in the atmosphere in the relatively small amounts now observed (around 2.4×10^{15} kg). The reaction of CO_2 with silicate-containing rocks to form Precambrian limestone suggests a possible moderating influence on its atmospheric concentration throughout geological time.

$$CaSiO_3 + CO_2 \rightarrow CaCO_3 + SiO_2$$

About ten percent of atmospheric CO_2 is taken up each year by photosynthesis. Of this, all except 0.05 percent is returned by respiration, almost entirely due to micro organisms. Carbon dioxide is also absorbed into the geochemical cycle, mostly as buried carbonate sediments.

Since the advent of the industrial revolution in the nineteenth century, the amount of CO_2 in the atmosphere has been increasing. Isotopic analysis shows that most of this has been due to fossil-fuel combustion; in recent years, the mass of carbon released to the atmosphere by this means has been more than ten times the estimated rate of natural removal into sediments. The large-scale destruction of tropical forests, which has accelerated greatly in recent years, is believed to exacerbate this effect by removing a temporary sink for CO_2.

The oceans have a large absorptive capacity for CO_2 owing to its reaction with carbonate:

$$H_2O + CO_2 + CO_3^{2-} \rightleftharpoons 2 HCO_3^-$$

There is about 600 times as much inorganic carbon in the oceans as in the atmosphere. However, efficient transfer takes place only into the topmost layer of the ocean, which contains only about one atmosphere equivalent of CO_2. Further uptake is limited by the very slow transport of water into the deep ocean, which takes around 1000 years. For this reason, the buffering effect of the oceans on atmospheric CO_2 is not very effective; it is estimated that only about ten percent of the additional CO_2 is taken up by the oceans.

Life as we know it on the Earth is entirely dependent on the tenuous layer of gas that clings to the surface of the globe, adding about 1% to its diameter and an insignificant amount to its total mass. And yet the atmosphere serves as the Earth's window and protective shield, as a

ATMOSPHERIC CO_2 CONCENTRATIONS

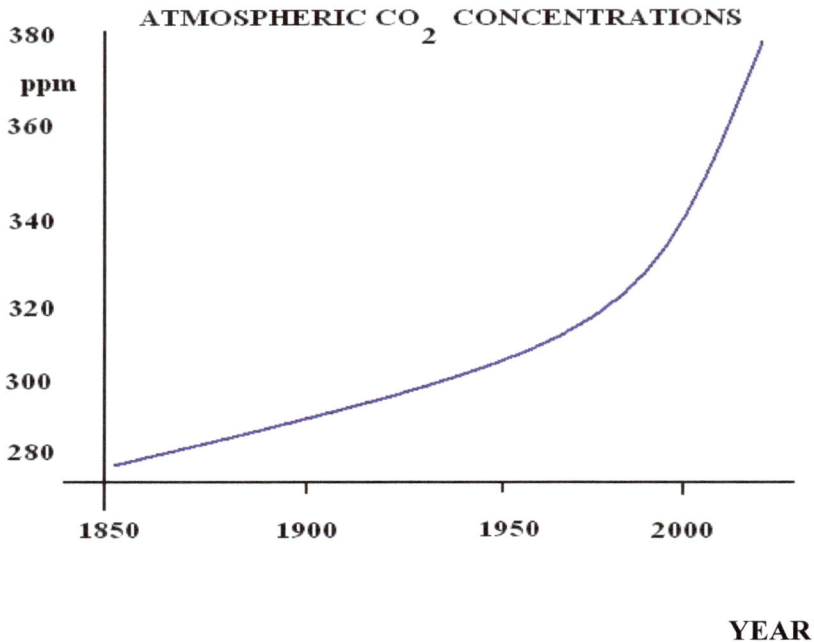

YEAR

medium for the transport of heat and water, and as source and sink for exchange of carbon, oxygen, and nitrogen with the biosphere. The atmosphere acts as a compressible fluid tied to the Earth by gravitation. As a receptor of solar energy and a thermal reservoir, it constitutes the working fluid of a heat engine that transports and redistributes matter and energy over the entire globe. The atmosphere is also a major temporary repository of a number of chemical elements that move in a cyclic manner between the hydrosphere, atmosphere, and the upper lithosphere. Finally, the atmosphere is a site for a large variety of complex photo-chemically initiated reactions involving both natural and human made substances.

Although we conveniently provide a distribution of compounds in the atmosphere, in reality the atmosphere is anything but uniform. Variations of temperature, pressure, and moisture content in the layers of air near the Earth's surface give rise to the dynamic effects we know as the weather.

Although the density of the atmosphere decreases without limit with increasing height, for most practical purposes one can roughly place its upper boundary at about 500 km. However, half the mass of the atmosphere lies within 5 km, and 99.99% within 80 km of the surface. The average atmospheric pressure at sea level is 101 kilo pascal. A 1-cm^2 cross section of the Earth's surface supports a column weighing 1030 g; the total mass of the atmosphere is about 5.3 x 10^{18} kg.

About 80% of the mass of the atmosphere resides in the first 10 km; this well-mixed region of fairly uniform composition is known as the *troposphere. The gases ozone, water vapour, and carbon dioxide are only minor components of the atmosphere, but they exert a huge effect on the Earth by absorbing radiation in the ranges indicated by the shading in the figure. Ozone in the upper atmosphere filters out the ultraviolet light below about 360 nm that is destructive of life. O_2, H_2O, CO_2 and CH_4 are "greenhouse" gases that trap some of the heat absorbed from the surface of the Earth and prevent it from re-radiating back into space.*

We think of gas molecules as moving about in a completely random manner, but the Earth's gravitational field causes downward motions to be very slightly favoured so that the molecules in any thin layer

ABSORPTION OF ENERGY BY GREENHOUSE GASES

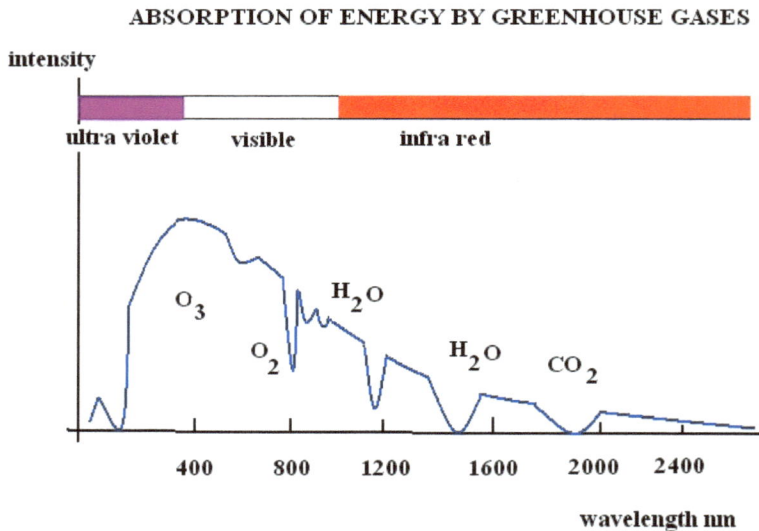

of the air collide more frequently with those in the layer below. This gives rise to a pressure gradient that is the most predictable and well known structural characteristic of the atmosphere. This gradient is described by an exponential law which predicts that the atmospheric pressure should decrease by 50% for every 6 km increase in altitude. This law also predicts that the composition of a gas mixture will change with altitude, the lower-molecular weight components being increasingly found at higher altitudes. However, this gravitational fractionation effect is completely obliterated below about 160 km owing to turbulence and wind. The atmosphere is divided vertically into several major regions which are distinguished by the sign of the temperature gradient. In the lowermost region, the troposphere, the temperature falls with increasing altitude. The major source of heat input into this part of the atmosphere is long-wave radiation from the Earth's surface, while the major loss is radiation into space.

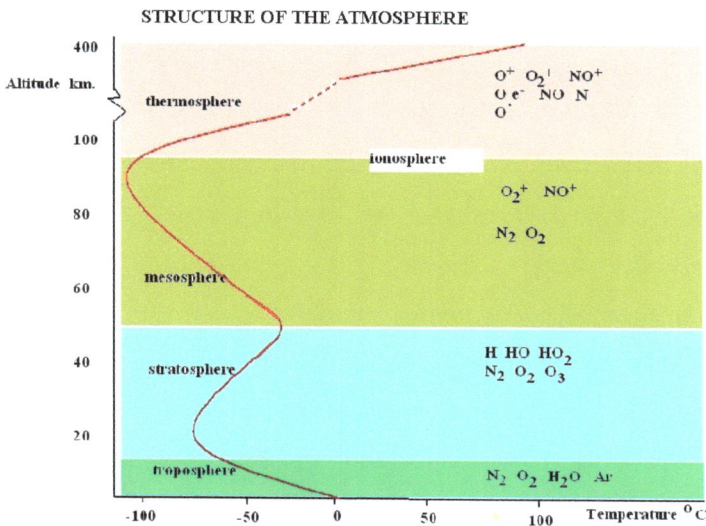

STRUCTURE OF THE ATMOSPHERE

46

At higher elevations the temperature begins to rise with altitude as we move into a region in which heat is produced by exothermic chemical reactions, mainly the decomposition of ozone that is formed photochemically from dioxygen in the stratosphere. At still higher elevations the ozone gives out and the temperature begins to drop; this is the mesosphere, which is finally replaced by the thermosphere which consists largely of plasma (gaseous ions). This outer section of the atmosphere which extends indefinitely to perhaps 2000 km is heated by absorption of intense ultra violet radiation from the Sun and also from the solar wind consisting of a continual rain of electrons, protons, and other particles emitted from the Sun's surface.

Major components		
nitrogen	N_2	78.08 %
oxygen	O_2	20.95 %
argon	Ar	0.93 %

With the exception of water vapour, whose atmospheric abundance varies from practically zero up to 4%, the fractions of the major atmospheric components N_2, O_2, and Ar are remarkably uniform below about 100 km. At greater heights, diffusion becomes the principal transport process, and the lighter gases become relatively more abundant. In addition, photochemical processes result in the formation of new species whose high reactivities would preclude their existence in significant concentrations at the higher pressures found at lower elevations. The atmospheric gases fall into three abundance categories: major, minor, and trace. Nitrogen, the most abundant component, has accumulated over time as a result of its chemical *inertness; a very small fraction of it passes into the other phases as a result of biological activity and natural fixation by lightning. It is believed that denitrifying bacteria in marine sediments may provide the major route for the return of N_2 to the atmosphere. Oxygen is almost entirely of biological origin, and cycles through the hydrosphere, the biosphere, and sedimentary rocks. Argon consists mainly of Ar^{40} which is a decay product of K^{40} in the mantle and crust. The most abundant of the minor gases aside from water vapour is carbon dioxide, about which more will be said below. Next in abundance are neon and helium. Helium is a decay product of radioactive elements in the Earth, but neon and the other inert gases are primordial, and have probably existed in their present relative abundances since the Earth's formation. Two of the minor gases, ozone and carbon monoxide, have abundances that vary with time and location. A variable abundance implies an imbalance between the rates of formation and removal. In the case of carbon monoxide, whose major source is anthropogenic (a small amount is produced by biological action), the*

Minor components		ppm
water	H_2O	0-4 %
carbon dioxide	CO_2	325
neon	Ne	18
helium	He	5
methane	CH_4	2
krypton	Kr	1
hydrogen	H_2	0.5
nitrous oxide	N_2O	0.3
carbon monoxide	CO	0.05-0.2

48

variance is probably due largely to localized differences in fuel consumption, particularly in internal combustion engines. The nature of the carbon monoxide sink (removal mechanism) is not entirely clear; it may be partly microbial.

Ozone is formed by the reaction of O_2 with oxygen atoms produced photochemically. As a consequence, the abundance of ozone varies with the time of day, the concentration of O_3 molecules and oxygen radicals from other sources (photochemical smog, for example), and particularly with altitude. The ozone concentration reaches a maximum of 12 ppm at 30 km.

The concentration of atmospheric carbon dioxide, while fairly uniform globally, is increasing at a rate of 0.2-0.7% per year as a result of fossil fuel burning. The present CO_2 content of the atmosphere is about 1.3 10^{17} kg. Most of the CO_2, however, is of natural origin, and represents the smallest part of the total carbonate reservoir that includes oceanic CO_2, HCO_3^-, and carbonate sediments. The latter contain about 600 times as much CO_2 as the atmosphere, and the oceans contain about 60 times as much. These relative amounts are controlled by

Trace components		ppb
ammonia	NH_3	4
nitric oxide	NO	1
sulphur dioxide	SO_2	1
hydrogen sulphide	H_2S	0.05

the rates of the reactions that interconvert the various forms of carbonate.

The surface conditions on the Earth are sensitively dependent on the atmospheric CO_2 concentration. This is due mainly to the strong infrared absorption of CO_2, which promotes the absorption and trapping of solar heat (see below). Since CO_2 acts as an acid in aqueous solution, the pH of the oceans is also dependent on the concentration of CO_2 in the atmosphere.

It is estimated that if only 1% of the carbonate presently in sediments were still in the atmosphere (as CO_2), the pH of the oceans would be 5.9, instead of the present 8.2.

$$H_2O + CO_2 \rightleftharpoons H_2CO_3$$

Energy Balance of the Atmosphere and Earth

The amount of energy or solar flux impinging on the outer part of the atmosphere is approximately 1400 watt m^{-2}. About 30% of this is reflected or scattered back into space by clouds, dust, and the atmospheric gas molecules themselves, and by the Earth's surface. About 19% of the radiation is absorbed by clouds or the atmosphere, mainly by $H2O$ and O_3 but not CO_2, leaving 51% of the incident energy available for absorption by the Earth's surface. If one takes into account the uneven illumination of the Earth's surface and the small flux of internal heat to the surface, the assumption of thermal equilibrium requires that the Earth emit about 240 watts m^{-2}. This corresponds to the power that would be emitted by a black body at − 18 °C, which is the average temperature of the atmosphere at an altitude of 5 km. The observed mean global surface temperature of the

Earth is 13°C, and is presumably the temperature required to maintain thermal equilibrium between the Earth and the atmosphere.

The energy radiated by the Earth has a longer wavelength than the incident radiation. Most gases absorb radiation in this range quite efficiently, including those gases such as CO_2 and N_2O that do not absorb the incident radiation. The energy absorbed by atmospheric gases is re-radiated in all directions. Some of it therefore escapes into space, but a portion returns to the Earth and is reabsorbed, thus raising its temperature. This is commonly called the greenhouse effect. If the amount of an infrared-absorbing gas such as carbon dioxide increases, a larger fraction of the incident solar radiation is trapped, and the mean temperature of the Earth will increase.

Any significant increase in the temperature of the oceans would increase the atmospheric concentrations of both water and CO_2, producing the possibility of a runaway process or feedback loop that would be catastrophic from a human perspective. Fossil fuel combustion and deforestation during the last two hundred years have increased the atmospheric CO_2 concentration by 25%, and this increase is continuing. The same combustion processes responsible for the increasing atmospheric CO_2 concentration also introduce considerable quantities of particulate materials into the upper atmosphere. The effect of these would be to scatter more of the incoming solar radiation, reducing the amount that reaches and heats the Earth's surface. The extent to which this process counteracts the greenhouse effect is still a matter of controversy; all that is known for certain is that the average temperature of the Earth is increasing.

Carbon dioxide is not the only atmospheric gas of manmade origin that can affect the heat balance of the Earth; other examples are SO_2 and N_2O. Nitrous oxide is of particular interest, since its abundance is fairly high, and is increasing at a rate of about 0.5% per year. It is produced mainly by bacteria, and much of the increase is probably connected with introduction of increased nitrate into the environment through agricultural fertilization and sewage disposal. Besides being a strong infrared absorber, N_2O is photochemically active, and can react with ozone. Any significant depletion of the ozone content of the upper atmosphere would permit more ultraviolet radiation to reach the Earth. This would have numerous deleterious effects on present life forms, as well as contributing to a temperature increase. The warming effect attributed to human additions of greenhouse gases to the atmosphere is estimated to be about 2 watt per m^2, or about 1.5% of the 150 watts per m^2 trapped by clouds and atmospheric gases. This is a relatively large perturbation compared to the maximum variation in solar output of 0.5 watts per m^2 that has been observed during the past century. Continuation of greenhouse gas emission at present levels for another century could increase the atmospheric warming effect by 6-8 watt per m^2.

A less-appreciated side effect of the increase in atmospheric carbon dioxide and of other plant nutrients such as nitrates may be a reduction in plant species diversity by selectively encouraging the growth of species which are ordinarily held in check by other species that are able to grow well with fewer nutrients. This effect, for which there is already some evidence, could be especially pronounced when *the competing species utilize different photosynthetic pathways that differ in their sensitivity to CO_2."*

The Earth and other planets are made up of the heavy elements (iron, silicon, nickel etc) as compared to the stars which contain mainly hydrogen and helium with only small amounts of these other elements. Only through many life cycles of stars have the heavier elements coalesced into vast clouds in space, to finally condense to make solar systems like our own. So have respect for the humble atom of iron (essential to the function of our blood cells). It most likely has a history of between some 10^{12} and 10^{14} years. The universe itself is comprised mainly of hydrogen so taking that perspective, planets and all that goes with them are somewhat of an oddity and rarity in the whole machine.

Coal contains between 45% and 55% carbon by mass. The amount of coal on the planet is (or rather was) around 6×10^{13} cubic metres or 8×10^{13} tonnes. When one considers that it took between 100 and 200 million years to create from ancient plants and forests it is little wonder that we are faced with considerable atmospheric problems due to burning the majority of this carbon in the space of a mere 300 years. (Tom Law predicts that it will all be gone by 2080)

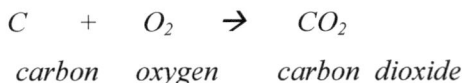

$$C \quad + \quad O_2 \quad \rightarrow \quad CO_2$$

carbon   oxygen   carbon dioxide

Natural gas from subterranean strata is a common fuel for both household and industrial use around the world. We might simplify its structure to butane which when fully combusted produces water and carbon dioxide:

$$2C_4H_{10\,(g)} \quad + \quad 13O_{2\,(g)} \quad \rightarrow \quad 8CO_{2\,(g)} \quad + \quad 10H_2O_{(l)}$$

butane $\qquad$ oxygen $\qquad$ carbon dioxide $\qquad$ water

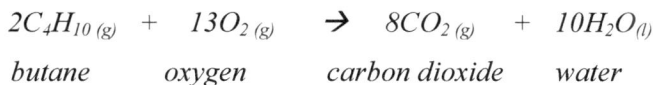

Hydrocarbon fuels from oil have been in use for only a hundred years. Of all the known resources of oil left on the Earth, Tom also predicts that it will be gone by 2073 at the current exponential rate of consumption. This then leaves plants and forest products as the last remaining but renewable resource of carbon and hydrocarbon compounds. I have no calculation on shale and shale oil here but it is relatively small compared to crude oil but nevertheless an important source. Australia has large untapped amounts of shale.

Limestone (calcium carbonate, $CaCO_3$) was laid down on the beds of shallow seas over a period of 400 million years. As this material is not a fuel, it is in not so great a demand. However it is still an important source of quicklime for the building industry and in the process releases carbon dioxide to the atmosphere.

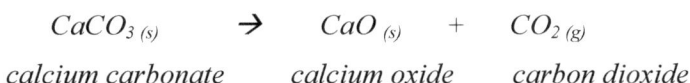

$$CaCO_{3\,(s)} \quad \rightarrow \quad CaO_{(s)} \quad + \quad CO_{2\,(g)}$$

calcium carbonate $\qquad$ calcium oxide $\qquad$ carbon dioxide

The production of metals usually starts with the ore which more often than not is an oxide, a carbonate or a sulphide. Thus again, in the reduction process, oxides of carbon and sulphur are released to the atmosphere, making use of atmospheric oxygen. An example might be the smelting of iron ore to make pure metallic iron:

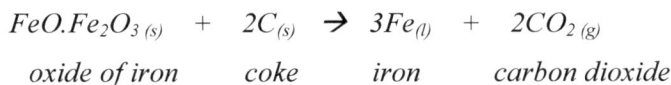

$$FeO.Fe_2O_{3\,(s)} \quad + \quad 2C_{(s)} \quad \rightarrow \quad 3Fe_{(l)} \quad + \quad 2CO_{2\,(g)}$$

oxide of iron $\qquad$ coke $\qquad$ iron $\qquad$ carbon dioxide

Limestone is also added to the blast furnace to assist in removing unwanted elements from the final product.

As a young science teacher I remember so many school excursions to the limestone deposits in East Gippsland, Victoria Australia, in search of fossils. To find a trilobite was extremely exciting. Many of our finds

were of primitive shellfish dating back in excess of 200 Mya. A sprinkling of the more fundamental Christians were not happy on my interpretations as they conflicted with their biblical understanding of the age of the Earth... and dinosaurs? They were just a figment of the imagination of scientists and movie script writers!

But the geological evidence is pretty overwhelming, with similar finds in many parts of the world. Whilst teaching recently in Changzhou, China, I was extremely excited by the museum pieces at the local park and botanical gardens lent by the Nanjing Institute of Geology and Mineral Resources. Not a big collection but some very nice specimens of ancient reptiles from the dinosaur period. Lastly, natural wood, peat, paper and other waste products are used by almost half the world's population to provide warmth in winter and a source of energy for everyday cooking of food. In some first world countries, wood chip is also being used as an energy source to produce

electricity, notably Germany and one or two Scandinavian countries.
Again we may simplify this chemical change to:

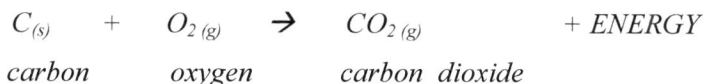

$$C_{(s)} \quad + \quad O_{2\,(g)} \quad \rightarrow \quad CO_{2\,(g)} \quad\quad + ENERGY$$

carbon   oxygen   carbon dioxide

Oh, I nearly forgot… there are you and me, all the fishes that live in the sea and all the cows, chickens, sheep, pigs etc that live on the land. And then there are the seaweed, the grass the cocoanut trees and all the plants upon the Earth… all things bright and beautiful! We are all farting out hydrogen sulphide and methane, and breathing out carbon dioxide. Oh and then there are all the rotting things at the garbage dump, the bogs and wetlands and in the soil. They are also producing methane, ammonia and some sulphur gases. It's all rather complex don't you think?

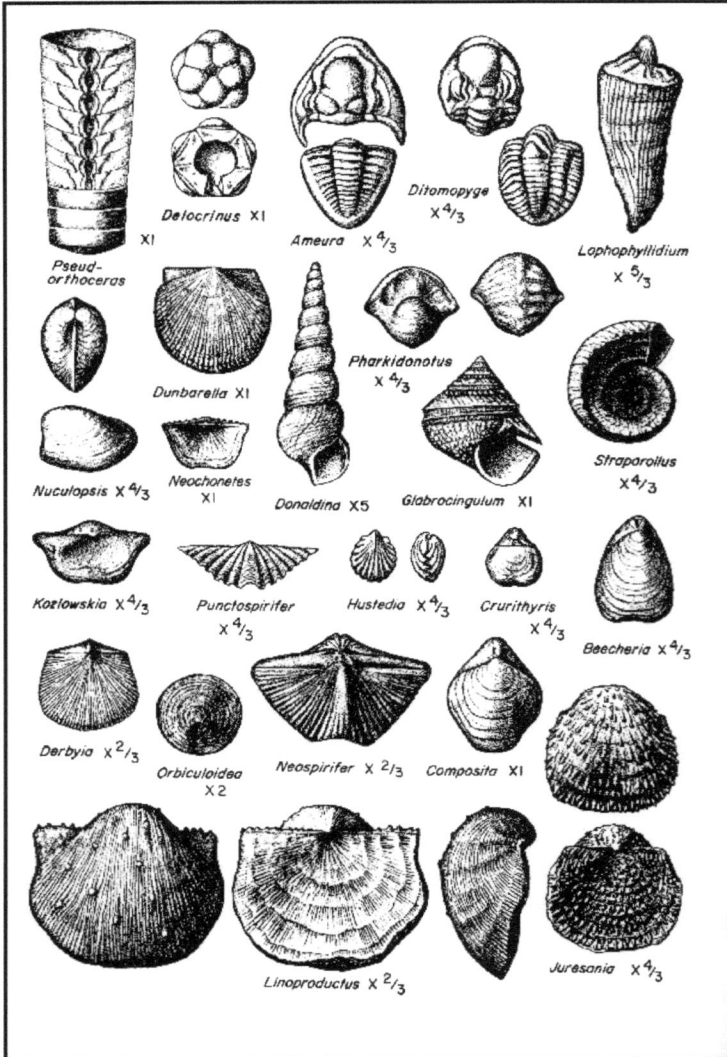

Pseudorthoceras X1 · *Delocrinus* X1 · *Ameura* X 4/3 · *Ditomopyge* X 4/3 · *Lophophyllidium* X 5/3 · *Dunbarella* X1 · *Pharkidonotus* X 4/3 · *Nuculopsis* X 4/3 · *Neochonetes* X1 · *Donaldina* X5 · *Glabrocingulum* X1 · *Straparollus* X 4/3 · *Kozlowskia* X 4/3 · *Punctospirifer* X 4/3 · *Hustedia* X 4/3 · *Crurithyris* X 4/3 · *Beecheria* X 4/3 · *Derbyia* X 2/3 · *Orbiculoidea* X2 · *Neospirifer* X 2/3 · *Composita* X1 · *Linoproductus* X 2/3 · *Juresania* X 4/3

But wait a minute, hang on… if all this is going on, where the hell is all the oxygen coming from that we need to breathe? Something's amiss!

57

The sun is shining, the sky is blue
mainly to oxygen it is due
the lotus pink, its flowery dew
and greeny leaves spread amidst
the watery pond held in my view.

Why of course… the plants and plankton (basic organisms in the top metre or so of the oceans) are doing the job by day. Yes they are photosynthesising! Most fourteen year olds have already seen the famous equation for photosynthesis:

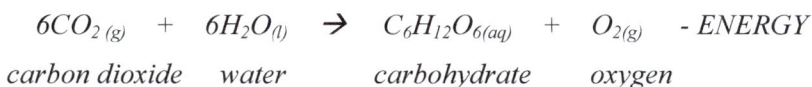

$$6CO_{2\,(g)} \quad + \quad 6H_2O_{(l)} \quad \rightarrow \quad C_6H_{12}O_{6(aq)} \quad + \quad O_{2(g)} \quad - ENERGY$$

carbon dioxide   water   carbohydrate   oxygen

and in those greeny leaves is chlorophyll, the magic substance that acts as the catalyst to combine these substances carbon dioxide and water along with the sun's rays to yield starches and sugars and most importantly, oxygen.. (I say magic, but I'm not so certain it is on the curriculum at Hoggwarts!) So here we have it, the whole system being buggered up by humans in the space of three hundred years and a bit more with all the responsibility of manufacturing oxygen placed on the plants and even the plants need to respire in order to live and grow.

Fortunately, they sequester more carbon in the form of structures (leaves, trunk, branches and fruit) than they expel as carbon dioxide in respiration. Thus we have an excess of oxygen production. This process has been going on for seven hundred million years and after a couple of hundred million years enough oxygen appeared in the atmosphere to support fish, reptiles, dinosaurs, birds and eventually mammals.

Equation for respiration:

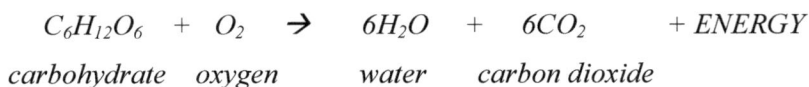

$$C_6H_{12}O_6 \quad + \quad O_2 \quad \rightarrow \quad 6H_2O \quad + \quad 6CO_2 \quad + ENERGY$$

carbohydrate   oxygen   water   carbon dioxide

Now the boffins around the world have come up with this solution: "If we can sequester carbon and carbon dioxide in a deep black hole somewhere, then all will be well and hunky dory; we may sleep easy and continue to breathe."

Sort of reminds me of a similar solution with nuclear waste!

(Sequester: to hide away, isolate, take possession of)

The hypocrisy and illogic of this solution is that:

(i)   *we continue to burn coal, gas and oil products, wood and anything we can get our grubby hands on and will continue to do this until there is nothing left to burn*

(ii)   *there is no proven safe way to sequester carbon and carbon dioxide other than by natural means i.e more forests and plants , clean seas, lakes and rivers*

Some experts have said it can be stored in deep rock layers or even under the sea bed (or under their own carpets and beds). Let us look at some of these what ifs:

What if some several trillions of tonnes of sequestered carbon dioxide were to suddenly break out due to continental drift, Earth tremor or similar natural occurrence? How well would we breathe then? It happened to the dinosaurs remember when a big icy comet landed. Only the littleys survived!

What if some entrepreneurial type came across an Aladdin's cave full of nice stored blocks of carbon? Is this not the same or even better than finding a coal mine?

No I don't think it is a lasting solution, only a temporary one whilst we run around shaking our heads looking for a better way. Mind you the comet idea is not so bad...

maybe its been done before! Would we shoot our big icy carbon dioxide snowball towards the sun or away? Would we receive complaints?

Dalton's chemical Symbols

oxygen   hydrogen   nitrogen   carbon   sulfur   phosphorous   gold

I think the next two chemical equations are as important as those for photosynthesis and respiration:

$$2H_2O \rightarrow 2H_2 + O_2 - ENERGY$$
$$2H_2 + O_2 \rightarrow 2H_2O + ENERGY$$

That is, the electrolysis of water to hydrogen and oxygen gases and the burning of hydrogen to form water. With the aid of the sun and solar panels this could be our future. An engine and fuel tank might be

completely sealed with just an endless cycle of these simple components to do mechanical work. There is no exhaust and certainly no production of any oxides of carbon, nitrogen or sulphur!

The nuclear fusion reaction:

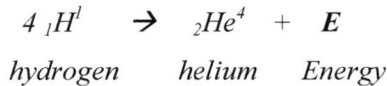

$$4 \ _1H^1 \quad \rightarrow \quad _2He^4 \quad + \quad E$$

$$hydrogen \qquad helium \qquad Energy$$

if it were technically feasible would also suddenly diminish the burden of CO_2 emissions. It is estimated that a single kilogram of hydrogen could theoretically produce 10^{14} Joule of heat energy in a fusion reaction. Of course we cannot do it just yet. But we must combine our efforts to achieve the dream of an endless supply of energy from such a minimal amount of fuel. Fission of uranium, as we know, produces unwanted radioactive waste that will be around for thousands of years. We need to stop this now.

The purist will argue that this fusion reaction already occurs in the nearest star... our sun (and probably deep inside the Earth). So why bother to even try to emulate it on the surface of the planet. It has *already been beaming down upon us and the planet for billions of years, perfecting an ideal system and balance. Perhaps we should stop meddling and revert to using the natural system sensibly. This, after all, is still science being applied intelligently is it not? So the idea of sequestering CO_2 is probably far-fetched and not really the ideal solution to the problem of excessive amounts of this gas. Of course I am not suggesting that we do nothing. If we are to change dramatically our means of energy production then it is a fair assumption that we must bear a period or transition. However, having said that, the period must not extend indefinitely as an excuse for 'business as usual'. Business as usual is the current direction we are taking and as said earlier will bring us to the very brink and within the space of a few decades. That is all the time we have left if we do not mend our ways!*

Examples of naturally sequestered carbon.

The biosphere comprises the various regions near the earth's surface that contain and are dynamically affected by the metabolic activity of the approximately 1.5 million animal species and 0.5 million plant species that are presently known and are still being discovered at a rate of about 10,000 per year. The biosphere is the youngest of the dynamical systems of the earth, having had its genesis about 2 billion years ago. It is also the one that has most profoundly affected the other major environmental systems, particularly the atmosphere and the hydrosphere.

The increase in the abundance of atmospheric oxygen from its initial value of essentially zero has without question been the most important single effect of life on earth, and the time scale of this increase parallels the development of life forms from their most primitive stages up to the appearance of the first land animals about 0.5 billion years ago.

There are many kinds of photosynthetic bacteria, but with one exception (cyanobacteria) they are incapable of using water as a source of hydrogen for reducing carbon dioxide. Instead, they consume hydrogen sulfide or other reduced sulfur compounds, organic molecules, or elemental hydrogen itself, excreting the reducing agent in an oxidized state. Green plants, cyanobacteria, green filamentous bacteria and the purple nonsulfur bacteria utilize glucose by respiration during periods of darkness, while the green sulfur bacteria and the purple sulfur bacteria are strictly anaerobic.

Present evidence suggests that blue-green algae, and possibly other primitive microbial forms of life, were flourishing 3 billion years ago. This brackets the origin of life to within one billion years; prior to 4 billion years ago, surface temperatures were probably above the melting point of iron, and there was no atmosphere nor hydrosphere. About 3.8 billion years ago, or one billion years after the Earth was formed, cooling had occurred to the point where rain was possible, and primitive warm, shallow oceans had formed. The atmosphere was anoxic and highly reducing, containing mainly CO_2, N_2, CO, H_2O, H_2S, traces of H_2, NH_3, CH_4, and less than 1% of the present amount of O_2, probably originating from the photolysis of water vapour. This oxygen would have been taken up quite rapidly by the many abundant oxidisable substances such as Fe(II), H_2S, and the like.

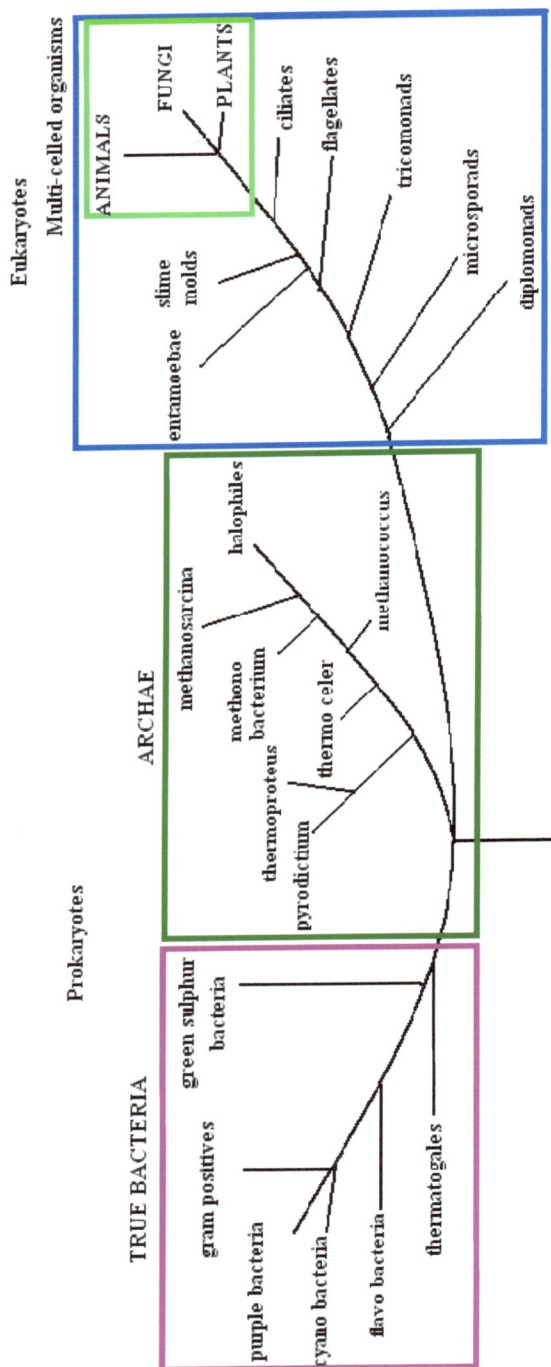

The Tree of Life

Eukaryotes

Multi-celled organisms

FUNGI

ANIMALS

PLANTS

ciliates

flagellates

tricomonads

microspords

diplomonads

entamoebae

slime molds

ARCHAE

halophiles

methanosarcina

methono bacterium

thermo celer

methanococcus

thermoproteus

pyrodictium

Prokaryotes

TRUE BACTERIA

green sulphur bacteria

gram positives

purple bacteria

cyano bacteria

flavo bacteria

thermatogales

65

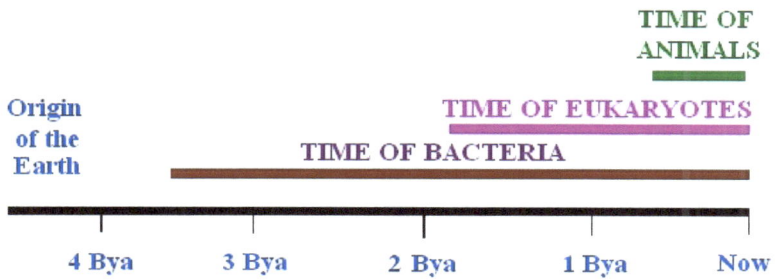

TIME OF ANIMALS

TIME OF EUKARYOTES

Origin of the Earth

TIME OF BACTERIA

| 4 Bya | 3 Bya | 2 Bya | 1 Bya | Now |

Life Time Line

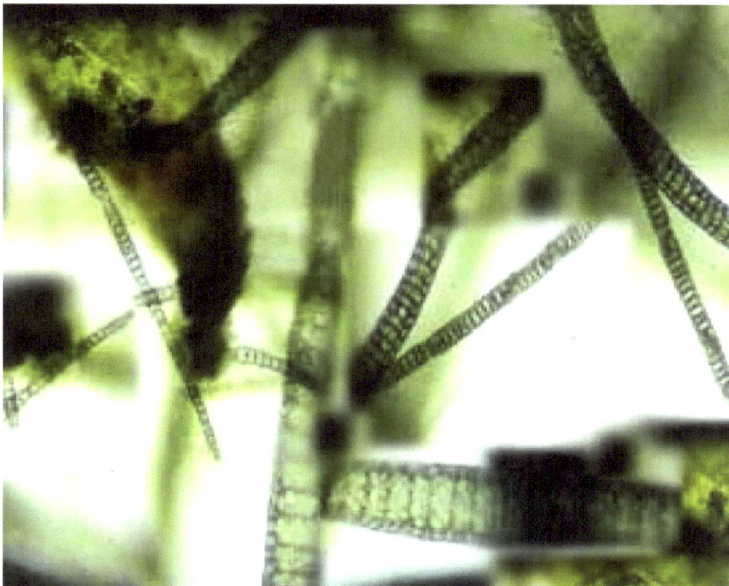

The fossil record that preserves the structural elements of organisms in sedimentary deposits has for some time provided a reasonably clear

66

picture of the evolution of life during the past 750,000 years. In more recent years, this record has been considerably extended, as improved techniques have made it possible to study the impressions made by single-celled microorganisms embedded in rock formations.

The main difficulty in studying fossil microorganisms extending back beyond a billion years is in establishing that the relatively simple structural forms one observes are truly biogenic. There are three major kinds of evidence for this.

- *Many of the most primitive life forms are still thriving, and these provide useful models with which some of the fossil forms can be compared.*
- *Carbon isotope ratios provide a second independent line of evidence for early life, or at least that of photosynthetic origin. In photosynthesis, $C^{12}O_2$ is taken up slightly more readily than is the heavier (and rarer) isotope $C^{13}O_2$; thus all but the very earliest life forms have left an isotopic fossil record even though the structural fossil may no longer be identifiable.*
- *A third evidence for early life is any indication of the presence of free oxygen in the local environment. Easily oxidisable species such as Fe(II) were very widely distributed on the primitive earth, and could not remain in contact with oxygen for very long without being oxidized. The oldest known formations of oxidized iron pyrite and of uranite are in sediments that were laid down between 2.0 and 2.3 billion years ago.*

If all three of these lines of evidence are present in samples that can be shown to be contemporaneous with the sediments in which they are found, then the argument for life is incontrovertible. One of the most

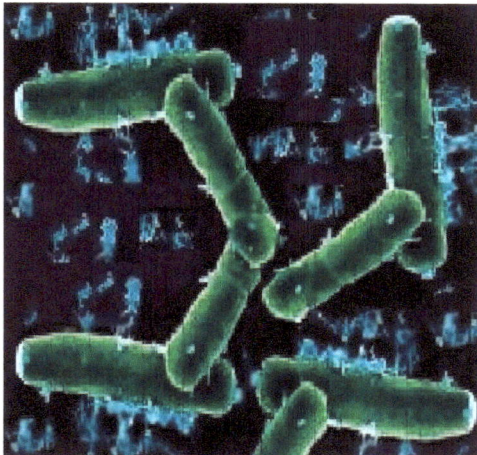

famous of these sites was discovered near Thunder Bay, Ontario in the early 1950's. The Gunflint Formation consists of an exposed layer of chert (largely silica) from which the overlying shale of the Canadian Shield had been removed. Microscopic examination of thin sections of this rock revealed a variety of microbial cell forms, including some resembling present freshwater blue-green algae. Also present in the Gunflint Deposits are the oldest known examples of metazoa, or organisms which display a clear differentiation into two or more types of cell. These deposits have been dated at 1.9-2.0 billion years.

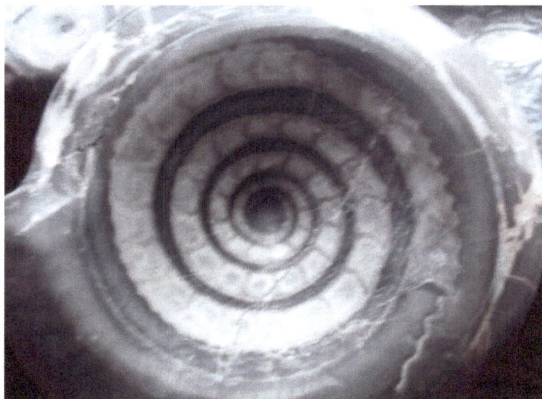

Western Australia has yielded fossil forms that are apparently 2.8 billion years old and other deposits in the same region contain structures resembling living blue-green algae. Other forms, heavily modified by chemical infiltration, bear some resemblance to a present iron bacterium, and are found in sediments laid down 3.5 billion years ago, but evidence that these fossils are contemporaneous with the sediments in which they are found is not convincing.

The oldest evidence of early life is the observed depletion of C^{13} in 3.8-billion year old rocks found in south-western Greenland. No laboratory experiment has yet succeeded in producing a self-replicating species that can be considered living, the mechanism by which this came about in nature must remain speculative(see previous chapter, The First Crawl). Infectious viruses have been made in the laboratory by simply mixing a variety of nucleotide precursors with a template nucleic acid and a replicase enzyme; the key to the creation of life is how to do the same thing without the template and the enzyme.

It has been estimated that about 50 genes are required in order to define the minimal biochemical and structural machinery that a hypothetical simplest possible cell would have.

Bacterial forms were likely the dominant form of life for several hundred million years. Eventually, due perhaps to the failing supply of H_2S, plants capable of mediating the photochemical extraction of hydrogen from water developed. This represented a large step in biochemical complexity; it takes 10 times as much energy to extract hydrogen from water than from hydrogen sulphide, but the supply is virtually limitless.

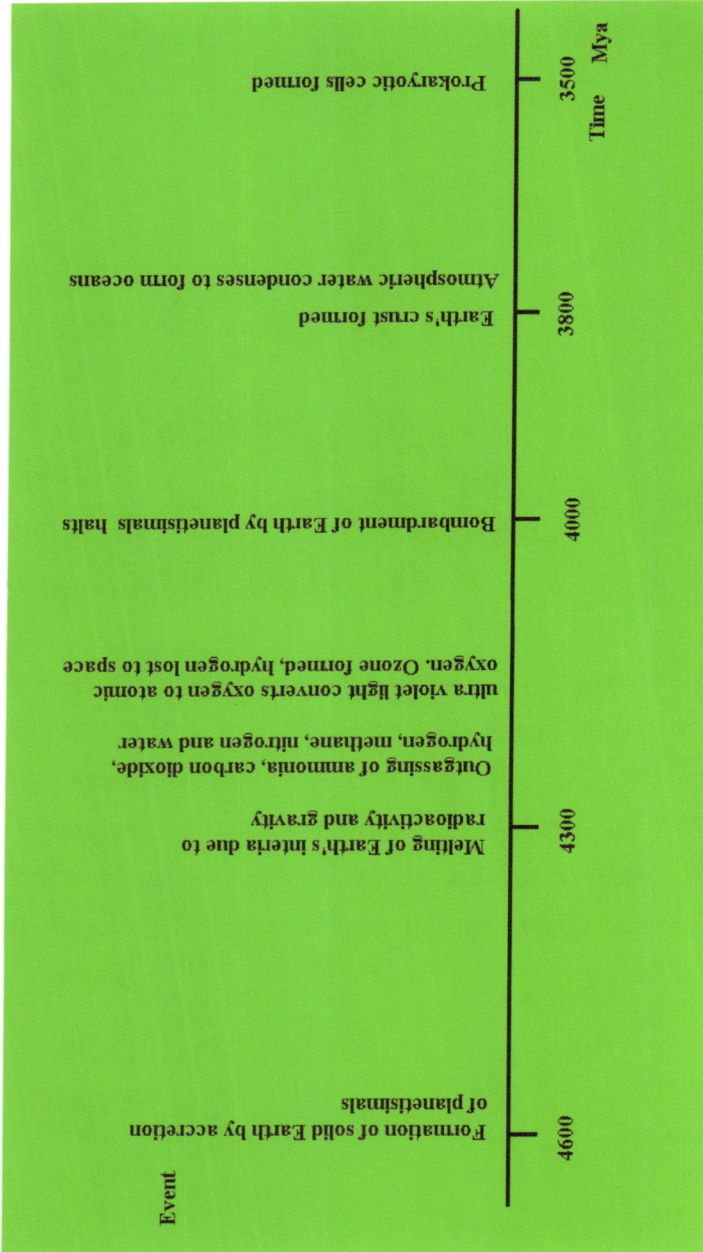

Time   Mya

Time (Mya)	Event
4600	Formation of solid Earth by accretion of planetesimals
4300	Melting of Earth's interia due to radioactivity and gravity
	Outgassing of ammonia, carbon dioxide, hydrogen, methane, nitrogen and water.
	ultra violet light converts oxygen to atomic oxygen, Ozone formed, hydrogen lost to space
4000	Bombardment of Earth by planetesimals halts
3800	Earth's crust formed
	Atmospheric water condenses to form oceans
3500	Prokaryotic cells formed

Event

71

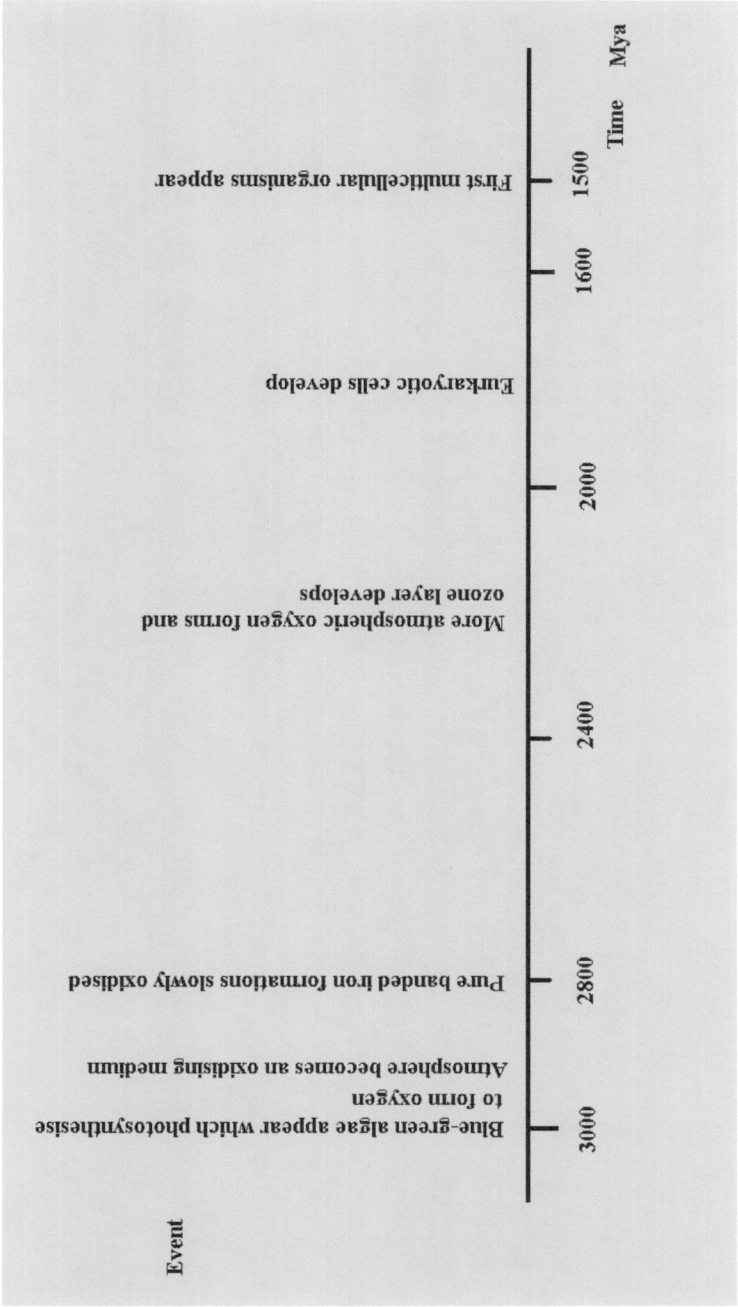

Time Mya

1500 — First multicellular organisms appear.

1600

Eukaryotic cells develop

2000

More atmospheric oxygen forms and ozone layer develops

2400

2800 — Pure banded iron formations slowly oxidised

Atmosphere becomes an oxidising medium

3000 — Blue-green algae appear which photosynthesise to form oxygen

Event

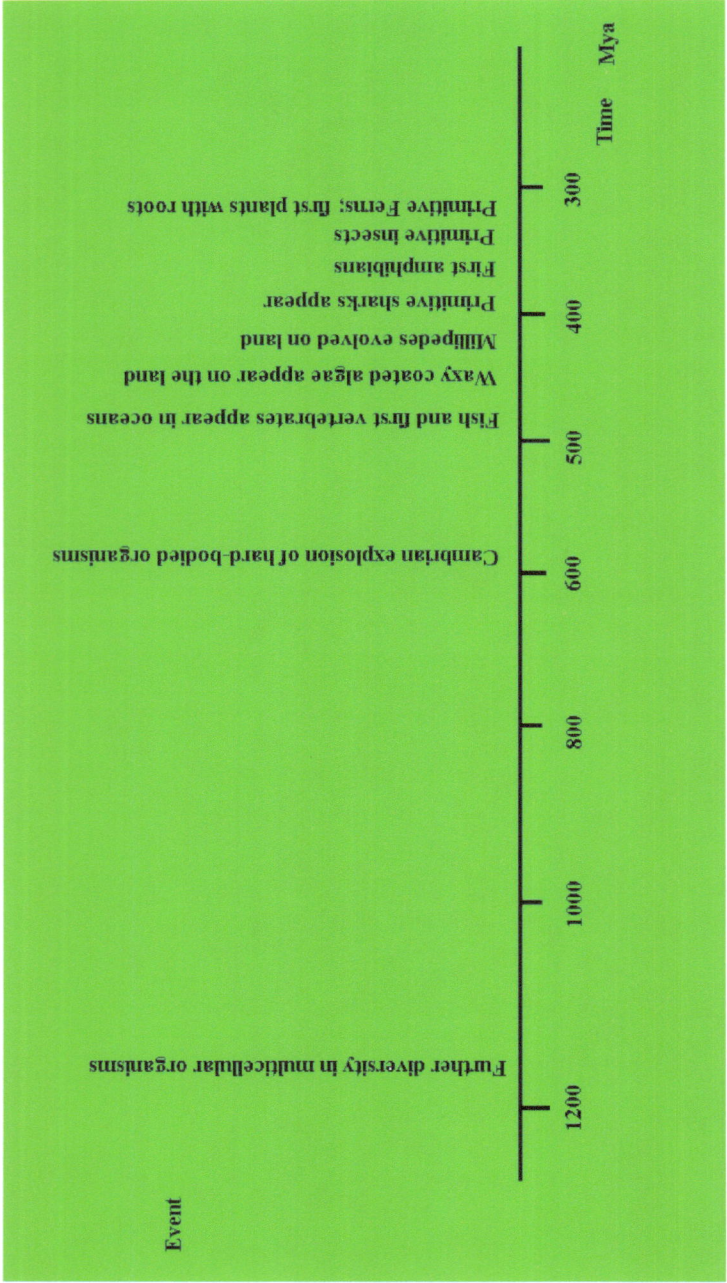

Time   Mya

300   Primitive Ferns; first plants with roots
       Primitive insects
       First amphibians
       Primitive sharks appear
400   Millipedes evolved on land
       Waxy coated algae appear on the land
       Fish and first vertebrates appear in oceans
500
600   Cambrian explosion of hard-bodied organisms
800
1000
1200   Further diversity in multicellular organisms

Event

73

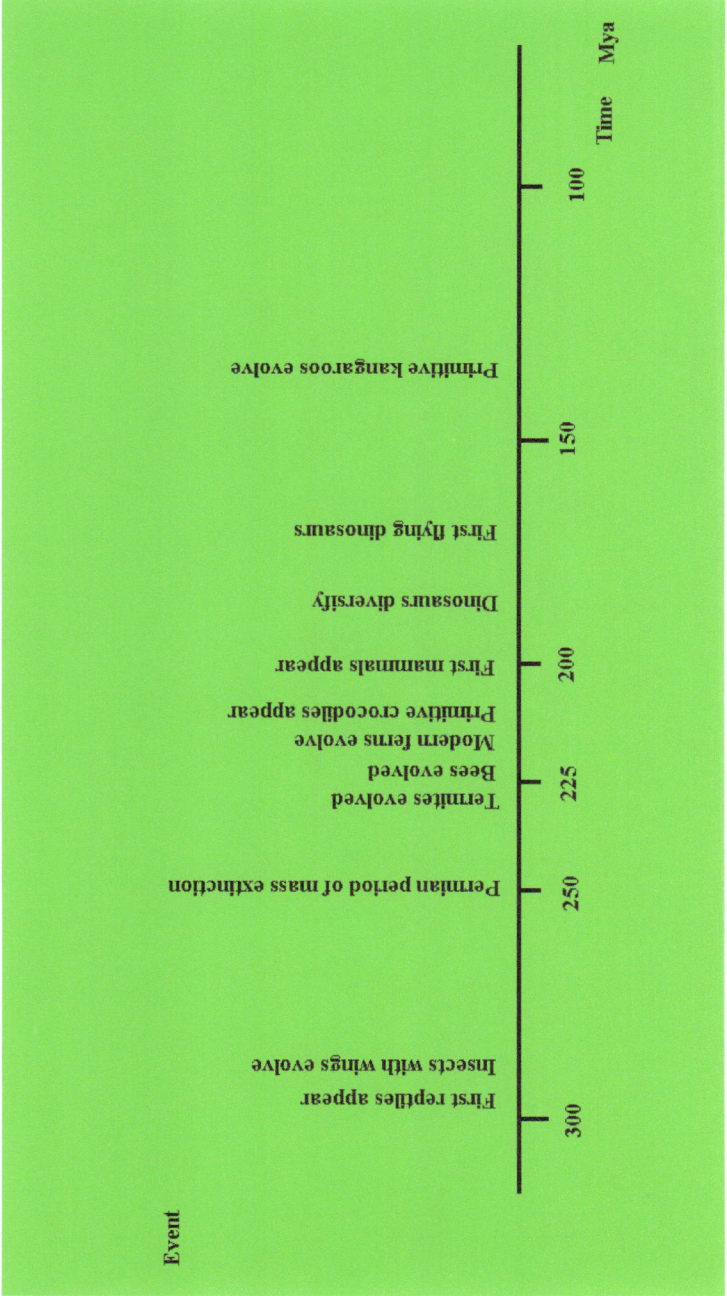

Event

- First reptiles appear
- Insects with wings evolve
- Permian period of mass extinction
- Termites evolved
- Bees evolved
- Modern ferns evolve
- Primitive crocodiles appear
- First mammals appear
- Dinosaurs diversify
- First flying dinosaurs
- Primitive kangaroos evolve

Time Mya: 300 — 250 — 225 — 200 — 150 — 100

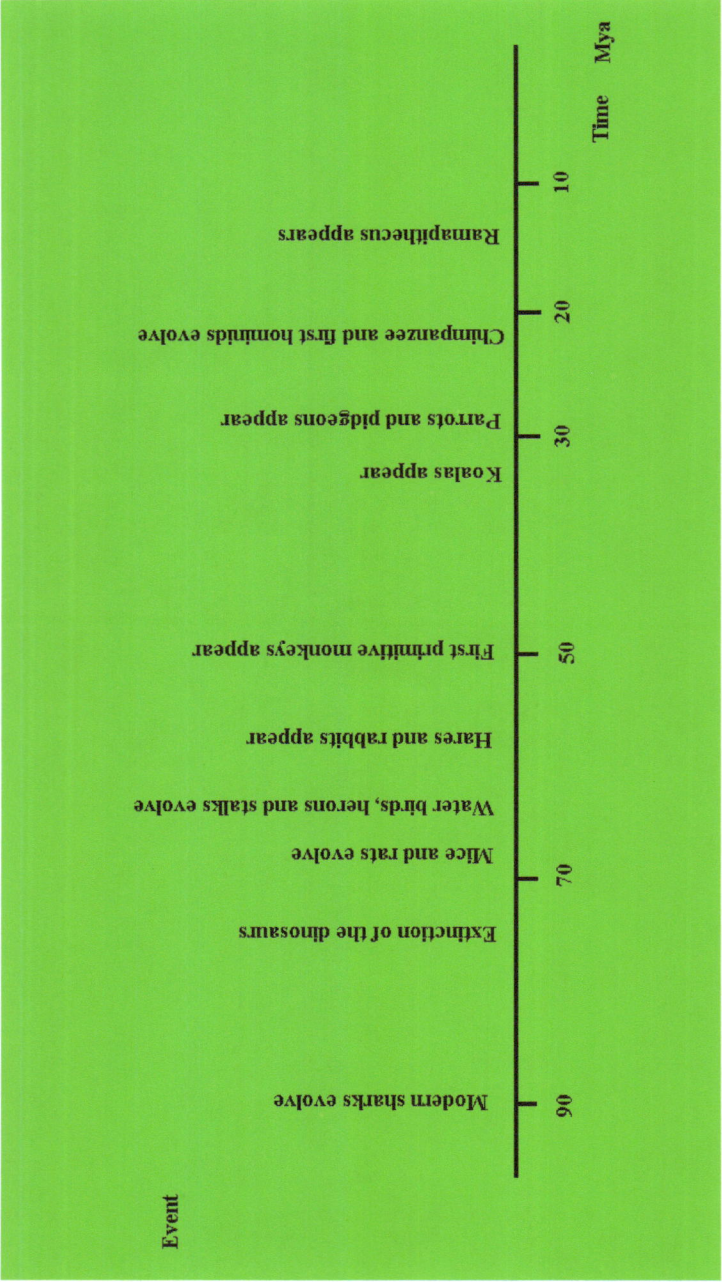

Event

Modern sharks evolve

Extinction of the dinosaurs

Mice and rats evolve

Water birds, herons and stalks evolve

Hares and rabbits appear

First primitive monkeys appear

Koalas appear

Parrots and pidgeons appear

Chimpanzee and first hominids evolve

Ramapithecus appears

Time   Mya

90   70   50   30   20   10

75

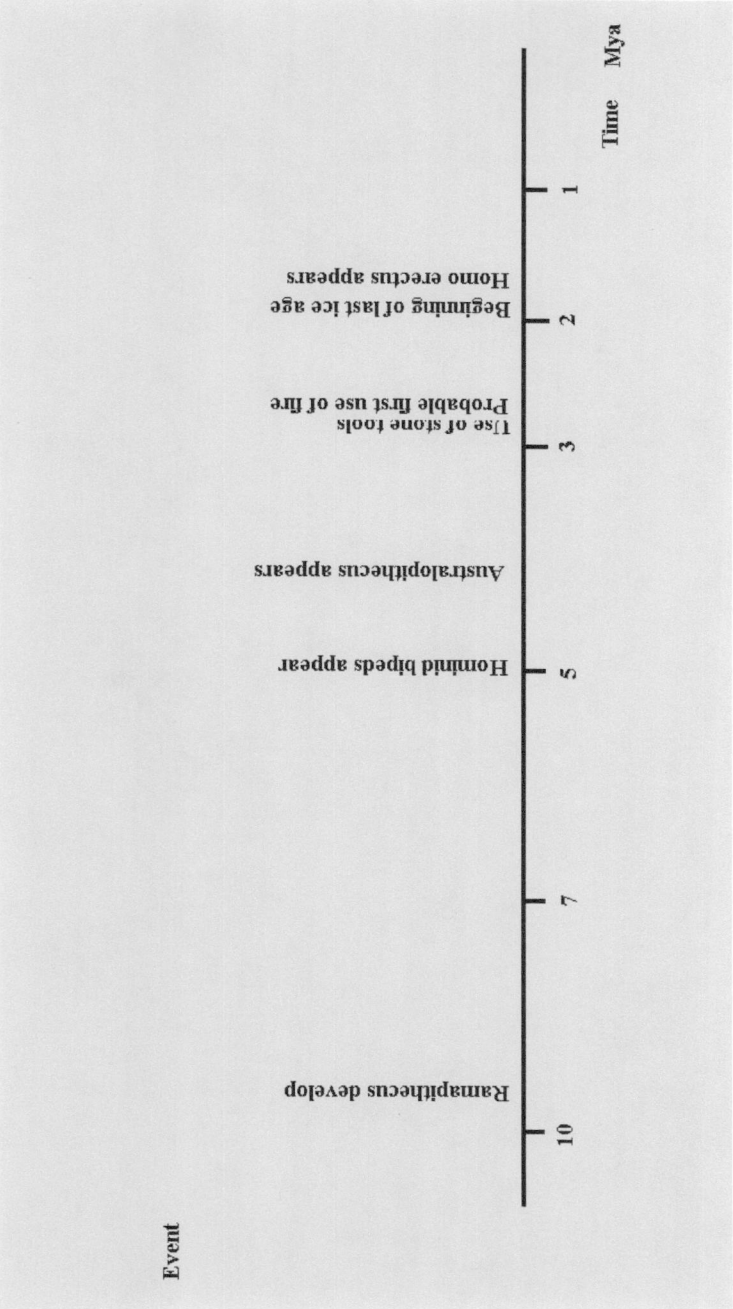

Event

Time   Mya

Time (Mya)	Event
1	
2	Homo erectus appears / Beginning of last ice age
3	Use of stone tools / Probable first use of fire
4	Australopithecus appears
5	Hominid bipeds appear
7	
10	Ramapithecus develop

Event

The last 500 thousand years

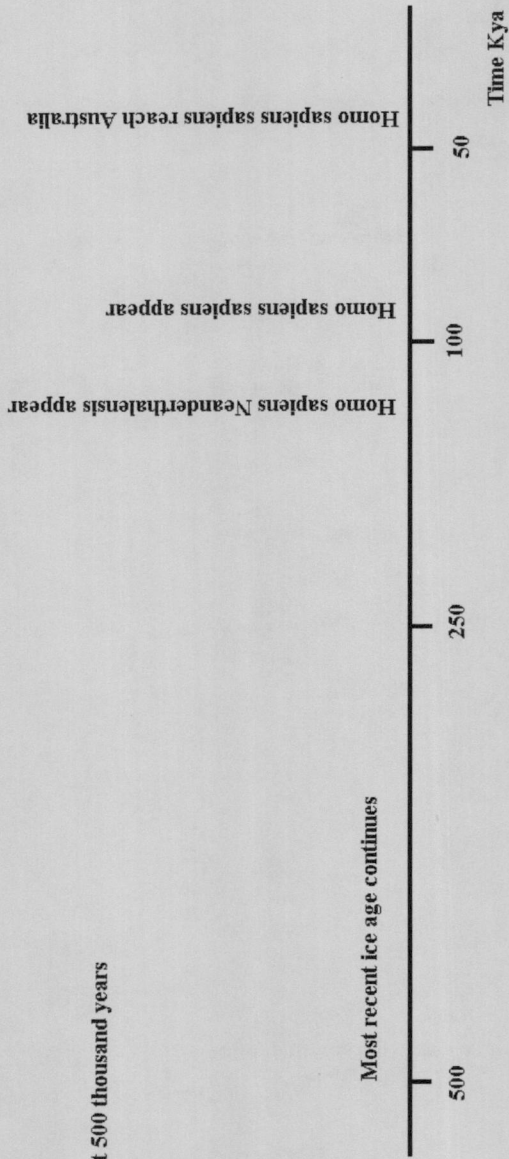

Most recent ice age continues

Homo sapiens Neanderthalensis appear

Homo sapiens appear

Homo sapiens reach Australia

Time Kya

500   250   100   50

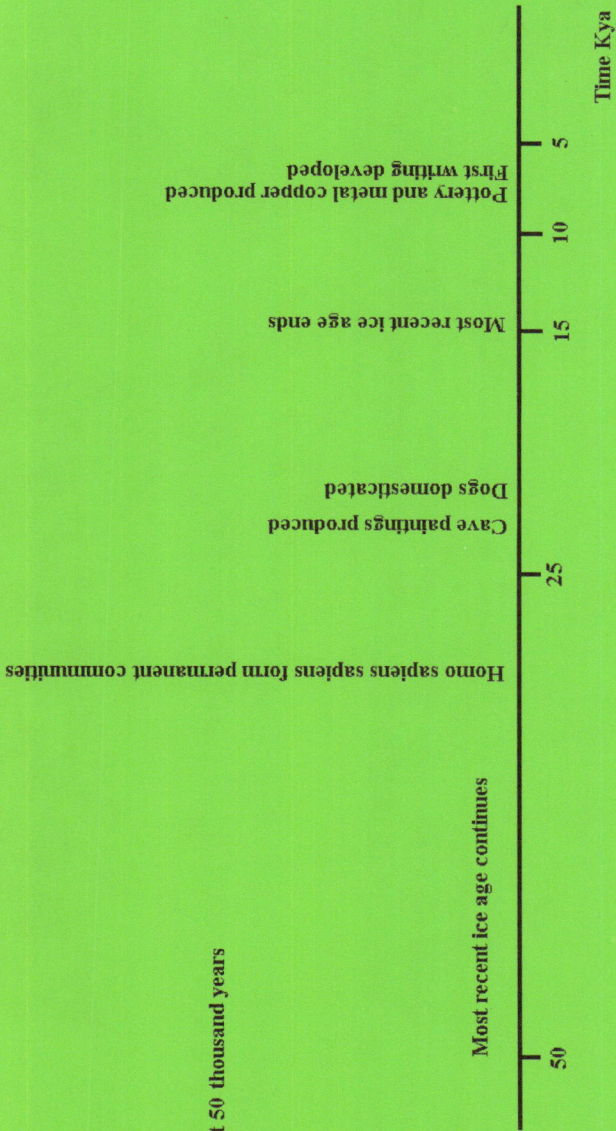

Event

The last 50 thousand years

Most recent ice age continues

Homo sapiens form permanent communities

Cave paintings produced
Dogs domesticated

Most recent ice age ends

Pottery and metal copper produced
First writing developed

Time Kya

50   25   15   10   5

78

Geological Timeline

Era – Period – Epoch			Time Mya
Archaeozoic (Archean) era			5000-1500
Proterozoic era			1500-545
Paleozoic era	Cambrian period		545-505
	Ordovician period		505-432
	Silurian period		432-410
	Devonian period		410-355
	Carboniferous period		355-290
	Permian period		290-250
Mesozoic era	Triassic period		250-205
	Jurassic period		205-135
	Cretaceous period		135-65
Cenozoic era "Recent Life"	Tertiary period	Palaeocene epoch	65-55
		Eocene epoch	55-38
		Oligocene epoch	38-26
		Miocene epoch	26-6
		Pliocene epoch	6-1.8
	Quarternary period	Pleistocene epoch	1.8-0.01
		(Lower Paleolithic)	0.90-0.25
		(Middle Paleolithic)	0.25-0.08
		(Upper Paleolithic)	0.08-0.01
		Holocene epoch	0.01-0

Bioregulation of the Atmosphere

The increase in the oxygen content of the atmosphere as a result of the development of the eukaryotic cell was discussed above. Why has the oxygen content levelled off at 21 percent? It is interesting to note that if the oxygen concentration in the atmosphere were only four percent higher, even damp vegetation, once ignited by lightning, would continue to burn, enveloping vast areas of the earth in a firestorm. Evidence for such a worldwide firestorm that may be related to the extinction of the dinosaurs has recently been discovered. The charcoal layers found in widely distributed sediments laid down about 65 million years ago are coincident with the iridium anomaly believed to be due to the collision of a large meteor with the Earth.

Regulation of the oxygen partial pressure is probably achieved by a balance between its production through photosynthesis and its consumption during oxidation of organic matter; the present steady state requires the burial of about 0.1% the carbon that is fixed annually, leaving one O_2 molecule in the air for each atom of carbon removed from the photosynthetic cycle. The large quantities of microbially-produced methane and N_2O also constitute important oxygen sinks; if methanogenic bacteria should suddenly cease to exist, the O_2 concentration would rise by 1% in about 12,000 years. This type of regulation implies a negative feedback mechanism, in which an increase in atmospheric oxygen would increase the activity of organisms capable of generating metabolic products that react with it.

Nitrous oxide, in addition to serving as an oxygen sink, might also be a factor in the regulation of the intensity of the ultraviolet component of sunlight. N_2O acts as a catalytic intermediate in the decomposition of stratospheric ozone, which shields the earth from excessive ultraviolet radiation.

Ammonia, another atmospheric gas, is produced by the biosphere in approximately the same quantities as methane, 10^9 tons per year, and at the expense of a considerable amount of metabolic energy. The function of NH_3 could well be to regulate the pH of the environment; in the absence of ammonia, the large amounts of SO_2 and HCl produced by volcanic action would reduce the pH of rain to about 3.

The fact that the atmospheric concentration of ammonia is only 10^{-8} times that of N_2 should not imply that this "trace" component plays a less significant role in the overall nitrogen cycle than does than N_2. In fact, the annual rates of production of the two gases are roughly the same; the much lower steady-state concentration of NH_3 is due to its faster turnover time.

As stable as the triply-bonded N_2 molecule is, there is a still more stable form of nitrogen: the hydrated nitrate ion. How is this stability consistent with the predominance of nitrogen in the atmosphere? The answer is that it is not: if it were not for nitrogen-fixing bacteria (powered directly or indirectly by the free energy of ATP captured from sunlight), the nitrogen content of the atmosphere would disappear to almost zero. This would raise the oxygen fraction to disastrously high levels, and the additional NO_3^- concentration would *increase the ionic strength and osmotic pressure of seawater to levels inconsistent with most forms of life.*

Bioregulation of the Oceans

The input of salts into the sea from streams and rivers is about 5.4 x 10^8 tons per year, into a total volume of about 1.2 x 10^9 km^3 yr^{-1} . Upwelling of juvenile water and hydrothermal action at oceanic ridges provide additional inputs of salts. With a few bizarre exceptions such as the brine shrimp and halophilic bacteria, 6 percent is about the maximum salinity level that organisms can tolerate. The internal salinities of cells must be maintained at much lower levels (around 1%) to prevent denaturation of proteins and other macromolecules whose conformations are dependent on electrostatic forces. At higher levels than this, the electrostatic interaction between the salt ions and the cell membrane destroys the integrity of the latter so that it can no longer pump out salt ions that leak in along the osmotic gradient. At the present rate of salt input, the oceans would have reached their present levels of salinity millions of years ago, and would by now have *an ionic strength far to high to support life, as is presently the case in the landlocked Dead Sea. The present average salinity of seawater is 3.4 percent. The salinity of blood, and of many other intra- and intercellular fluids in animals, is about 0.8 percent. If we assume that the first organisms were approximately in osmotic equilibrium with seawater, then our body fluids might represent "fossilized" seawater as it existed at the time our predecessors moved out of the sea and onto the land.*

By what processes is salt removed from the oceans in order to maintain a steady-state salinity? This remains one of the major open questions of chemical oceanography. There are a number of answers, mostly based on strictly inorganic processes, but none is adequately supported by available evidence. For example, Na^+ and Mg^{2+} ions could adsorb to particulate debris as it drops to the seafloor, and become incorporated into sediments. The requirement for charge conservation might be met by the involvement of negatively charged silicate and hydroxyaluminum ions. Another possible mechanism might be the burial of salt beds formed by evaporation in shallow, isolated arms of the sea, such as the Persian Gulf. Extensive underground salt deposits are certainly found on most continents, but it is difficult to see how this very slow mechanism could have led to a constant and steady salinity over shorter periods of highly variable climatic conditions.

The possibility of biological control of oceanic salinity starts with the observation that about half of the Earth's biomass resides in the sea, and that a significant fraction of this consists of diatoms and other organisms that build skeletons of silica. When these organisms die, they sink to the bottom of the sea and add about 300 million tons of silica to sedimentary rocks annually. It is for this reason that the upper levels of the sea are undersaturated in silica, and that the ratio of silica to salt in dead salt lakes is much higher than in the ocean.

These facts could constitute a basis for a biological control of the silica content of seawater; any link between silica and salt could lead to the control of the latter substance as well. For example, the salt ions might adsorb onto the silica skeletons, and be carried down with them; if the growth of these silica-containing organisms is itself dependent on salinity, we would have our negative feedback mechanism. The continual build-up of biogenic sedimentary deposits on the ocean floor might possibly deform the thin oceanic crust by its weight, and cause local heating by its insulating properties. This could conceivably lead to volcanic action and the formation of new land mass, thus linking the lithosphere into Gaia.

Three   **Coal, Oil and Limestone**

Coal

What is coal and where does it come from? How much have we used and how much is left? Coal is believed to have been formed from the debris of ancient forests which compacted over hundreds of millions of years. We can find fossils of insects and small vertebrae as well as the various parts of plants if we analyse coal. Coal deposits are distributed across the world and the resource has been used as a fuel for thousands of years. Only since the beginnings of the industrial revolution in Europe (Circa 1750 onwards) has coal been burned in large quantities, particularly for the production of iron and later for the production of electricity in power stations. With the advance of industrialisation across the globe, coal has been vital to the economies and lifeblood of nations. In 2004 some 6 billion tonnes of coal were burned, contributing to 21 billion tonnes tonnes of carbon dioxide to the atmosphere. Much of this is absorbed back into the environment i.e into plants and the oceans. But analysis of carbon dioxide in Antarctic ice samples shows that at no point

Atmospheric Carbon Dioxide Levels

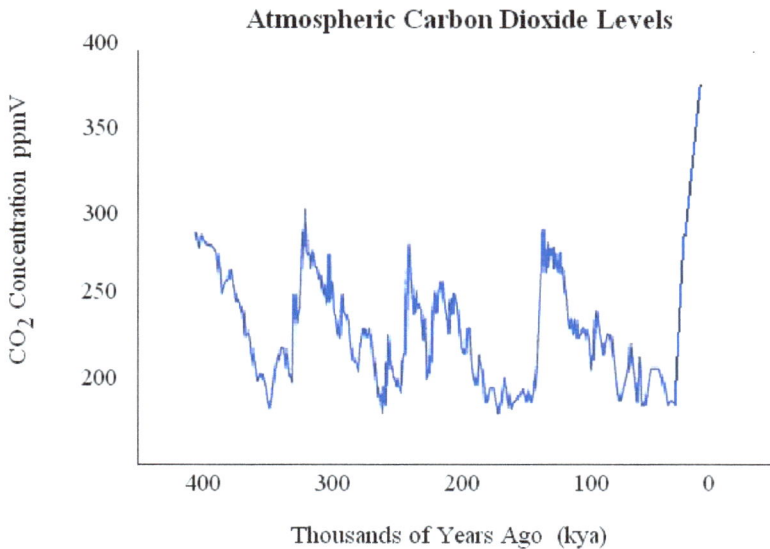

during the past 650,000 years did levels approach today's carbon dioxide concentrations of around 380 parts per million.* The Intergovernmental Panel on Climate Change (IPCC) projects that atmospheric carbon dioxide levels could reach 450-550 ppm by 2050, possibly resulting in higher temperatures and rising sea levels. There is fear that climate change could create millions of displaced persons from their homes by rising oceans, increasingly catastrophic weather, and expanding deserts.

[This assumes no sharp peaks in the data over extremely short time periods such as one hundred years or less which would be impossible to detect]*

Since 1750 we have used up approximately 300 billion tones of coal. Mining geologists assure us that we still have one trillion tones remaining, sufficient to last two or more hundred years. But is this so? At an exponential average increase of just 1.8 % per annum we see that all coal will be used up before 2080.

Total World Coal Demand

MegaTonne

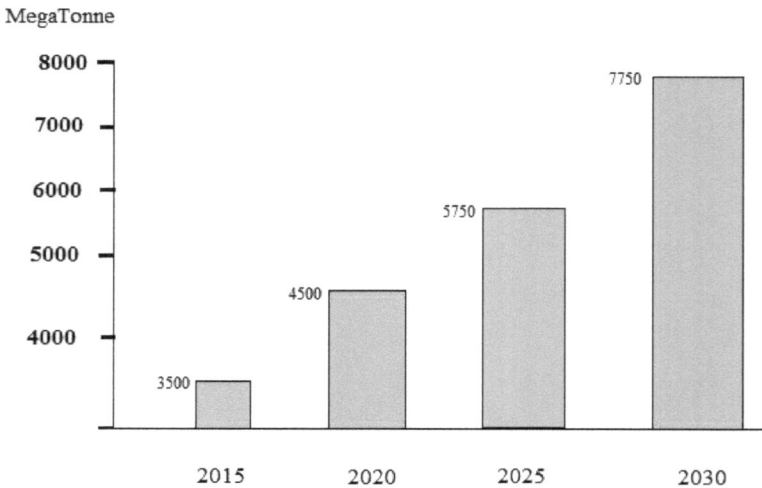

The bar chart shows Total World Coal Demand in MegaTonne: 2015 = 3500, 2020 = 4500, 2025 = 5750, 2030 = 7750.

The new giant economies of China and India, whilst reducing the percentage of their total power production attributed to coal (see table) in fact will still demand an increasing supply due to the rapid increase in their overall power requirements. Thus, even with nuclear power expansion and a scramble for friendlier alternatives such as *solar and wind, the lion's share of power production will continue to rely on coal until it is all gone. This has horrific implications for the atmosphere and the well being of humankind and the planet. The future outpouring of carbon dioxide from coal fired power stations will reach alarming levels exacerbating the current trend and forcing dramatic changes in weather and climate generally around the world. It is predicted by some scientists that a sudden collapse of the atmosphere as we know it is possible once CO_2 levels reach some critical value. The effects on our environment are both unpredictable and not understood. Initially we may experience meteoric rises in surface temperature but dramatic drop in upper atmosphere temperatures.*

Coal Consumption China/India 2004-2054 (million tonne)

China	Year	Total Requirement (coal equivalent)	Coal Consumption	Percentage of Total Requirement
	2005	3150	2050	65%
	2025	4250	2350	55%
	2055	8250	3150	38%
	Total		**125000**	
India	2005	1050	500	48%
	2025	7250	4000	55%
	2055	12000	3950	33%
	Total		**185000**	

Table 4.1

The table 4.1 describes incredible increase in total energy demand for both China and India. Each of these nations is increasing its nuclear electricity component dramatically and there is a drive for alternative energy supply such as wind, solar and geothermal. Will these last three make any impression on the projected figures for coal produced

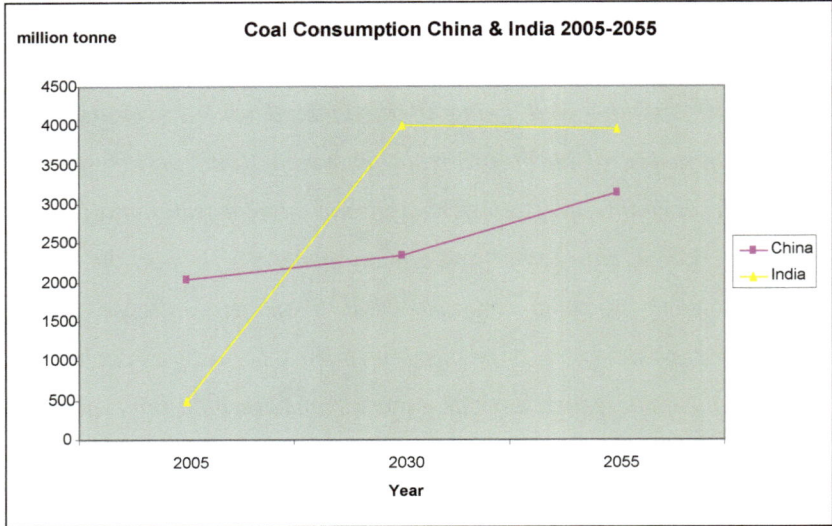

Coal Consumption China & India 2005-2055

Graph 4.1

electricity? One would hope so.

The best case scenario for world coal consumption is if it can be pegged back to 1.5 % mean exponential increase per year. However this only gives us until 2086, a mere additional six years until the resource has been entirely swallowed by a hungry world. Even if it were critically pegged back to 1% exponential annual increase in world demand (seen as impossible by realist economists), coal would be gone by around 2150. I am repeating here what I have already said in my earlier book "Nuclear Islam..." that the world faces a catastrophe in the late 2070's when considering human population growth, technological advance by current third world countries together with the sudden disappearance of energy resources. It is a mathematical fact as certain as the engineer's prediction that the

Titanic will founder! This is serious stuff you might exclaim. Yes, and particularly so for our grandchildren.

But even if this amount of coal is available after say 2030, what will the atmosphere be like by then? What will the mean temperature of the Earth and her oceans each be by then? What human catastrophe awaits us? Will there be sufficient oxygen for us to breathe?

The Last Straw

I cannot stress more greatly at this juncture how important it now is for each and every nation to plan and determine its future in terms of population, food, energy and natural resources. There is no longer any room for the sceptics. Science and mathematics has shut them up. It is certain that without an about turn and swing to renewable energy sources on a grand scale, humanity is doomed not to survive until the end of the current century. This will happen without any other cause such as a major war. Alternative renewable energy sources must fully be in place for 90% or better of the world demand by around 2050 if not earlier. I need to qualify this statement.

If the nations continue on the current path, we are going to see serious repercussions long before these cut-off dates like 2050 or 2080. Let us make the fair assumption that between 1750 and 2005, the world saw around 240000 million tonnes of coal burned. Even with current planning and intention of reducing the share of coal in producing power for home and industry worldwide, we will have outpoured CO_2 from an equivalent tonnage again by 2033. What sort of a world do you think the people of that time will inherit? It will be totally inhospitable. We will have contributed a further 880000 million tonnes of carbon dioxide gas to the atmosphere. And what are the *likely consequences of such a dramatic increase in just 28 years? I predict that the oceans, plants and insects would not be able to absorb such a vast amount of the gas. In all likelihood the concentration of CO2 will have risen to between 1500 and 2000 ppm. The mean surface temperature of the land will have risen by between 8° and 14° Celsius making much of the planet uninhabitable. Sea levels will have risen by around 18 metres and the mean ocean temperature will have risen by 4.5° Celsius. The disruption to normal daily life and all the structures, systems and mechanisms of modern society will have collapsed into sudden chaos. Millions of people will die in the confusion. A possible immediate follow up to this chaos on the surface will be a sudden cooling in the troposphere to temperatures not seen since the onslaught of the last ice age. In other words there will be a sudden inversion of the normality of the atmosphere viewed as a whole. (See various possible scenarios below)*

ENVIRONMENTAL BALANCE

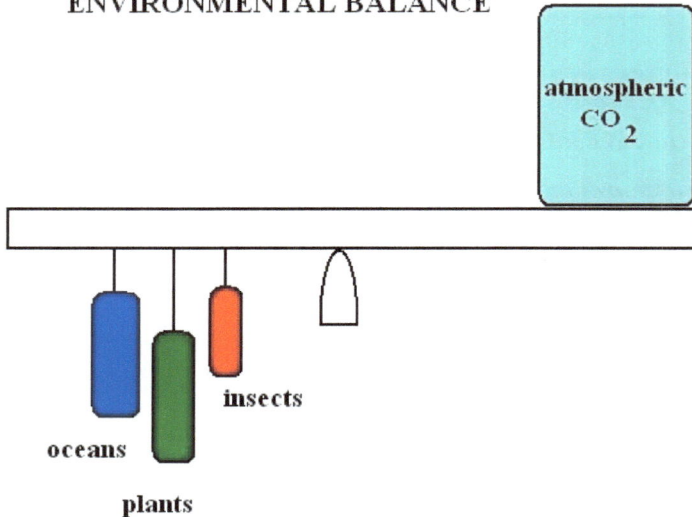

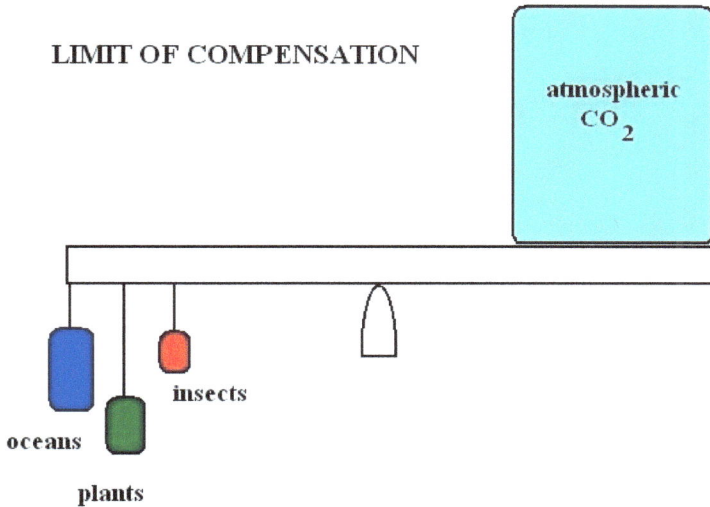

LIMIT OF COMPENSATION

atmospheric CO_2

oceans

plants

insects

Scenario 1   7.5^0 C

This describes one possibility for an increase in the mean temperature of the surface of the Earth by 7.5^0 C:

As a consequence of the massive increase in evaporated moisture from the land and the sea, we will experience unprecedented rain, snow and hail raining down as massive chunks of ice perhaps up to a metre in diameter, destroying anything upon which they fall, including sturdy houses and buildings. This ice will take a sizeable proportion of CO_2 with it. Such a sudden dump of ice and snow over the whole world will now have the opposite effect of the high CO_2 concentrations. Sunlight and heat will now be radiated more efficiently into space accelerating an ice age. The flip will not take thousands of years but will descend within the space of a decade. My prediction for this outrageous

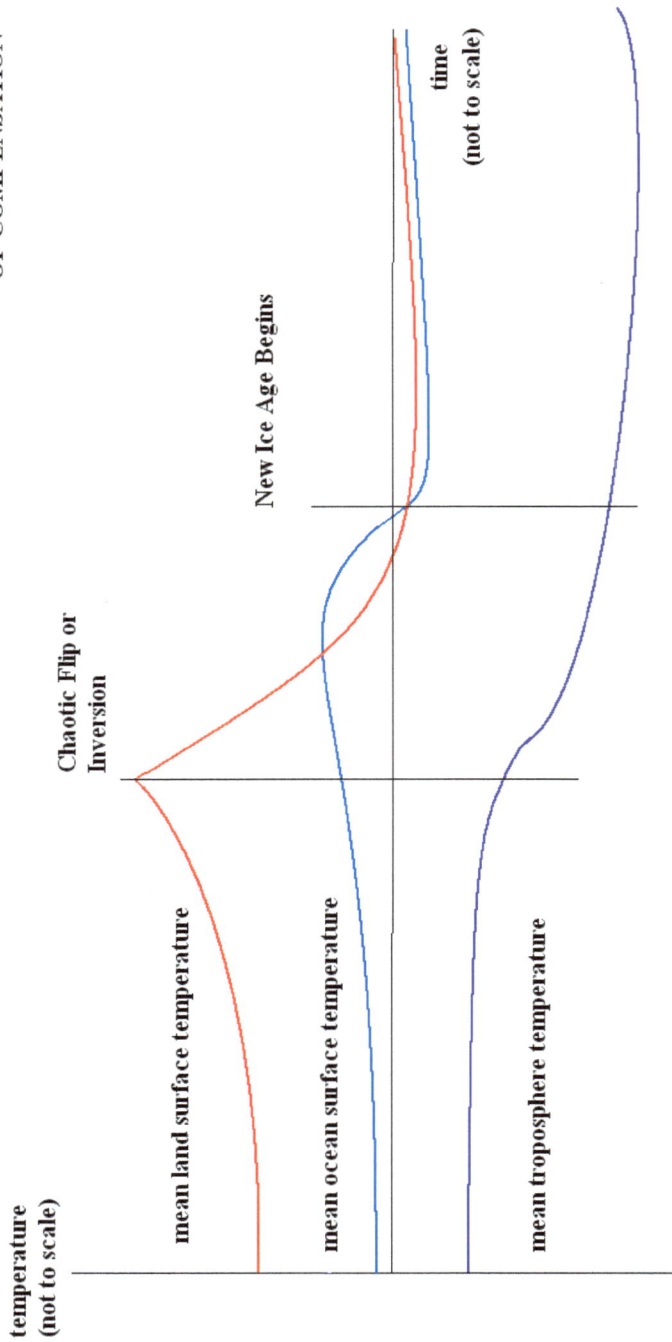

CHAOTIC INVERSION OF LAND, SEA AND TROPOSPHERE TEMPERATURES AT THE CO_2 LIMIT
OF COMPENSATION

temperature
(not to scale)

Chaotic Flip or
Inversion

New Ice Age Begins

mean land surface temperature

mean ocean surface temperature

mean troposphere temperature

time
(not to scale)

scenario is for some time between 2023 and 2033. The time interval between the chaotic flip or inversion and the beginning of a new ice age I am saying might be sudden… as little as ten years. The above scenario has not yet been predicted by computer models of climate and none such has been published in the scientific journals. 2033… just 25 years from the time of writing, April 2009!

Scenario 2   7.5⁰ C

Another alternative scenario of extreme reaction to an increase in the mean temperature of the surface of the Earth by 7.5⁰ C is runaway global warming due to a saturated atmosphere with water. As seen in the figure below for scenario 2, the water molecule is also a greenhouse gas. At some limit (which I suggest here is a mean increase of 7.5⁰ C temperature rise) a feedback loop will occur (i.e a recursive input of heat to the atmosphere) due to ever increasing amounts of water from the oceans. This may result in a rapid increase of land surface temperatures to those of a pressure cooker killing all living organisms. Further, water vapour reaching the upper atmosphere will boil off into space at an alarming rate. Within the period of one or two hundred years the Earth will resemble the planet Mars with little atmosphere and just a little water frozen at its poles.

Scenario 3   7.5⁰ C

According to recent research both the Arctic and the Antarctic were ice-free from 100 mya until about 40 mya. Between 80 mya and 60 mya dinosaurs roamed Antarctica where foliage was plentiful. Sea levels were approximately seventy metres above those of today. Ferns, cycads and conifers grew on the Antarctic continent with surrounding

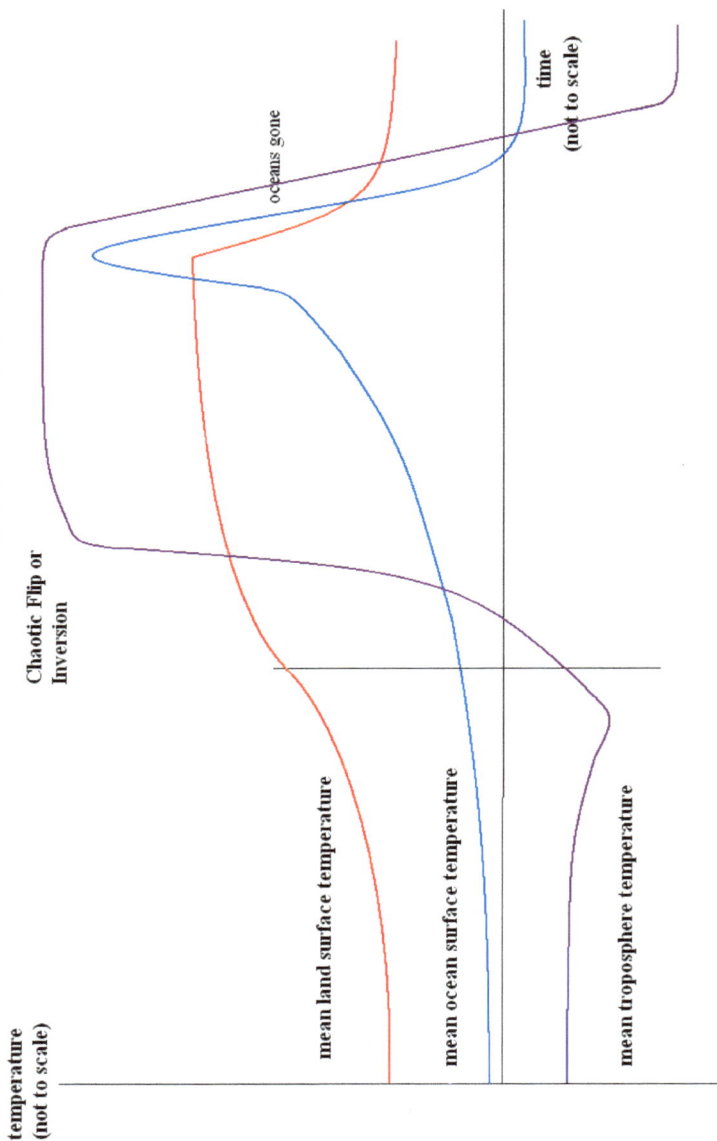

CHAOTIC INVERSION OF LAND, SEA AND TROPOSPHERE TEMPERATURES AT THE CO_2 LIMIT
OF COMPENSATION - SCENARIO 2

temperature
(not to scale)

Chaotic Flip or
Inversion

oceans gone

time
(not to scale)

mean land surface temperature

mean ocean surface temperature

mean troposphere temperature

95

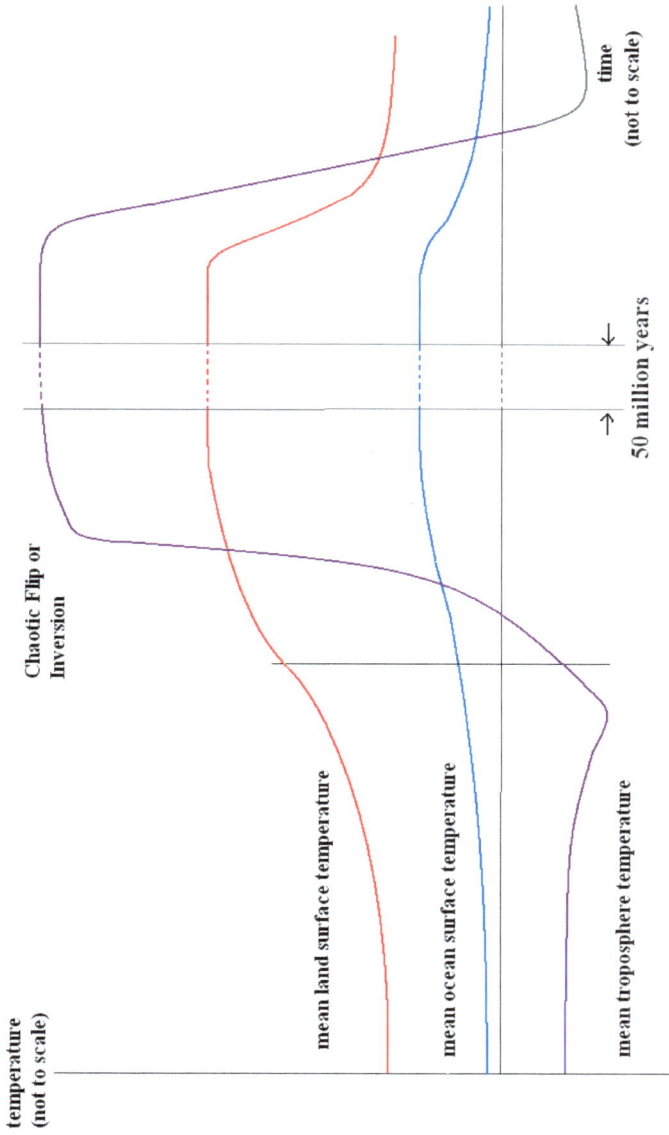

CHAOTIC INVERSION OF LAND, SEA AND TROPOSPHERE TEMPERATURES AT THE CO_2 LIMIT

OF COMPENSATION - SCENARIO 3

temperature
(not to scale)

Chaotic Flip or
Inversion

mean land surface temperature

mean ocean surface temperature

mean troposphere temperature

50 million years

time
(not to scale)

seas between 17 ^0C and 19 ^0C. Freshwater turtles swam in the surface waters of the Arctic region where ferns also grew. Mean land temperatures varied between 40 ^0C and 50 ^0C. CO_2 levels are estimated to have been some 12 times the immediate pre-industrial level.

What initiated this hothouse phase of some sixty million years? It might best be described as the 'runaway greenhouse effect'; but what caused such a sudden rise in the CO_2 level? One explanation is the onslaught of volcanic action on a grand scale interacting with large areas of limestone formations. This would have released a large and unprecedented volume of CO_2 gas into the atmosphere raising the temperatures of land, sea and troposphere to trigger the second stage water greenhouse effect. Instead of all water boiling away into space, however, the brief chaotic system reached a new equilibrium that lasted sixty million years with balmy weather at the poles, scorching inland temperatures and rain that would make even the biblical 40 days and 40 nights seem paltry by comparison. But which of these do we face in the later half of the twenty-first century?

Liquid Gold- Oil

We love our modern world. Particularly we love to drive our car or automobile as they say in America. Engines in cars run on gasoline. They must also be lubricated with oil and grease. The tires are made from rubber and much of the materials making the seats and dashboard are made from plastics. The wiring is also sheathed in plastic. The road we drive upon is mostly made from bitumen. All these materials are derivatives from crude oil, pumped from deep under ground or from beneath the floor of the ocean. Now here is a simple question. Can you see the day when every Asian, African and South American family owns at least one car driven by petroleum or LPG? Well I do not. A simple calculation will immediately tell you that if all the families described did in fact own a car, each vehicle would be able to drive for approximately four weeks (about 1400 km in total) before all the known reserves of oil on this planet would be gone. A sobering thought! Bye-bye Miss American Pie!!

So where and how was oil and gas formed?

Oil has been created continually throughout much of the Earth's history and is being formed in some parts of the Earth today. Almost all oil and gas comes from tiny decayed plants, algae, and bacteria. At certain times in the Earth's history conditions for oil formation have been particularly favourable. For instance, oil from the North Sea is mainly found in rocks that formed during the Jurassic period, some 150 Mya, long before humans appeared on the Earth. During this time shallow seas and swampy areas were rich in microscopic plants and animals. When these died they slowly sank to the bottom forming thick layers of organic material. This in turn became covered in layers of mud that trapped this organic material.

Oil and gas were formed by the anaerobic decay of organic material in conditions of increased temperature and pressure. Anaerobic means in a condition with the absence of oxygen.

These layers of mud prevented air from reaching the organic material. Without air, the organic material is unable to decay in the same way as organic material rots away in a compost heap for example. As the layers of mud grew in thickness, they pushed down on the organic

material with increasing pressure. The temperature of the organic

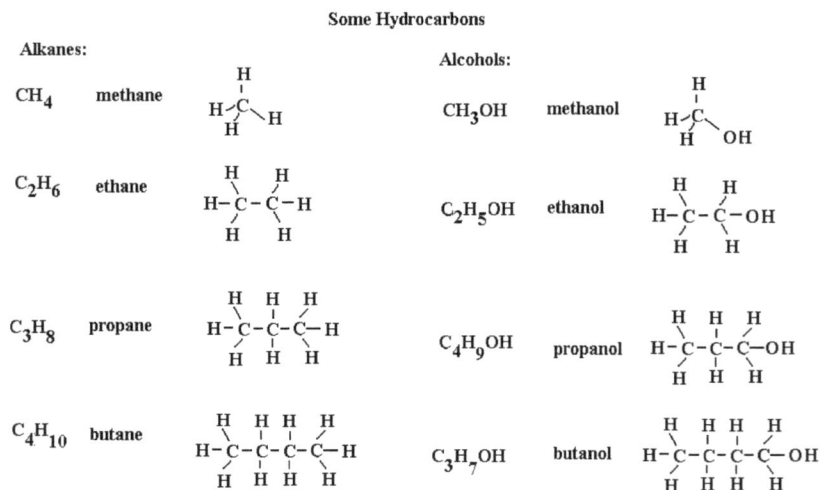

Some Hydrocarbons

Alkanes:

CH_4 methane

C_2H_6 ethane

C_3H_8 propane

C_4H_{10} butane

Alcohols:

CH_3OH methanol

C_2H_5OH ethanol

C_4H_9OH propanol

C_3H_7OH butanol

material was also increased as it was heated by other processes going

on inside the Earth. Very slowly, increasing temperature, pressure

and anaerobic bacteria started acting on the organic material. As this

happened the material was slowly cooked and chemically altered. In

this way, that initial energy first given to the plants by the sun is

transferred and the organic matter converted to the raw materials

crude oil and gas. However, the oil forms first, then as the

temperature and pressure increase at greater depths, gas also begins

to form. The temperature gradient within the Earth's crust increases

with depth so that sediments and plant material they contain warm up

as they become buried under more sediment. Increasing heat and

pressure first causes the buried algae, bacteria, spores and leaf skin to

join their wax, fat and oil to form dark specks called kerogen.

The cellulose and woody part of plants are converted to coal and woody kerogen. Rocks containing sufficient organic substances to generate oil and gas in this way are known as source rocks. When the source rock starts to generate oil or gas it is said to be mature.

Some heavier Hydrocarbons

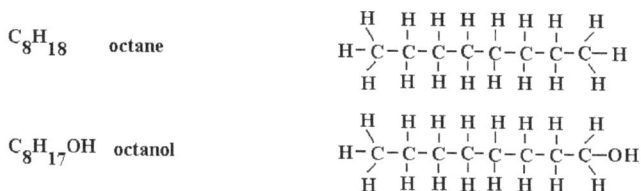

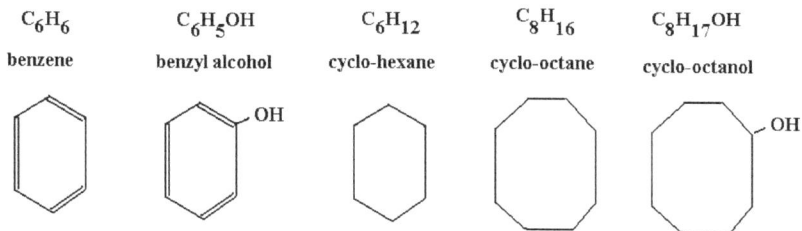

C_8H_{18} octane

$C_8H_{17}OH$ octanol

C_6H_6	C_6H_5OH	C_6H_{12}	C_8H_{16}	$C_8H_{17}OH$
benzene	benzyl alcohol	cyclo-hexane	cyclo-octane	cyclo-octanol

As the source rock gets hotter, chains of hydrocarbon chemicals use this heat energy to break away from the kerogen to form waxy and viscous heavy oil. At greater depth where conditions of higher temperature and pressure exist, the chains of hydrocarbons become shorter and break away to give light oil and gas. Most North Sea Oil and that of Bass Strait in Southern Australia is the valuable light oil. Gas from these sites is predominantly methane. Oil and gas are called 'hydrocarbons' because they mostly contain molecules of the elements hydrogen and carbon.

Crude oil is a complex mixture of hydrocarbons with small amounts of other chemical compounds that contain elements such as sulphur, nitrogen and oxygen. Traces of other elements, such as sulphur and nitrogen present in the decaying organic material give rise to small quantities of other compounds in crude oil.

Hydrocarbon molecules come in a variety of shapes and sizes such as straight-chain, branched chain or cyclic. This is one of the things that make them so valuable because it allows them to be used in so many different ways. By various chemical techniques such as fractional distillation, a variety of components may be separated. This is done at an oil refinery. From these fractions, a whole range of products such as petrol, diesel, aviation fuel, bitumen and the gas ethane may be formed. Ethene is the starting point for so many other useful products such as polythene, plastics and polymers.

The Story of Limestone

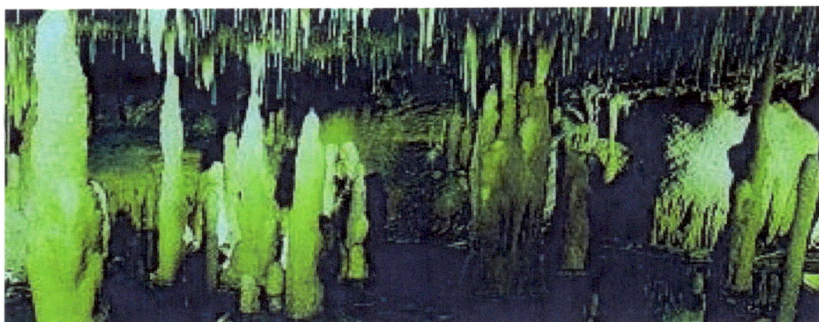

Limestone is a rock made up of mostly calcite, a mineral form of calcium carbonate ($CaCO_3$). Most fresh water and seawater contain dissolved calcium carbonate. Limestone is formed when the calcium carbonate crystallizes out of solution. There are generally two groups of limestone.

The first group includes limestone that has formed without the aid of organisms such as the microscopic foraminifera. Evaporating water forces this kind of limestone out of solution. A "lime" mud is deposited on the bottom of the sea. This white mud slowly hardens into a light-coloured limestone with a fine grain and even layers which is why limestone is generally softer than other stone materials. Limestone can also form on land when water evaporates leaving calcium carbonate to form a crust.

The second group of limestone is formed through the aide of organisms. Many aquatic organisms such as oysters, clams, snails, corals, and sea urchins as well as fish have shells or bones constructed of calcium carbonate. When these animals die, their shells and bones are broken up by waves into a mud which forms layers.

Limestone is a sedimentary rock and constitutes about ten percent of all sedimentary rocks. It may form inorganically or by biochemical processes. There are many types of limestone because of the variety of conditions under which it is produced.

Coral reefs are examples of limestone produced in the form of the skeletons of the coral invertebrate animals. Calcium-carbonate secreting algae live with the corals and help to cement the structures together. Large limestone deposits from ancient reefs are found inland. Chalk is another form of biochemically produced limestone. Chalk is a soft, porous rock made up of the skeletal parts of *microscopic marine organisms. An example is the cliffs of Dover, England. Coquina is the name given to limestone in the form of poorly cemented shells and shell fragments.*

Limestone usually forms in shallow water less than 20 m deep and thus also provides important geological information on the variation in sea level in the past. It frequently also contains the mineral dolomite which is calcium magnesium carbonate: $CaMg(CO_3)_2$

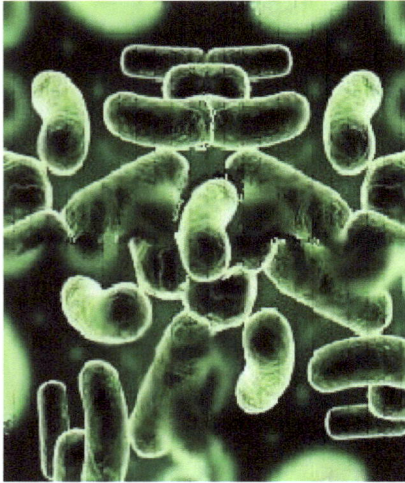

The colour of limestone is due to impurities such as sand, clay, iron oxides and hydroxides, and organic materials. When calcium carbonate precipitates, it can form two different minerals— calcite and aragonite. Calcite and aragonite are polymorphs, i.e they have the same chemical composition, but the atoms are arranged differently in the crystal structure.

Lastly, the solubility of calcite and dolomite, and the rate at which solution occurs, are dependent on at least eight factors: amount of carbon dioxide in solution, pH, oxidation of organic matter, temperature, pressure, concentration of added salts, rate of solution flow, and the degree of solution mixing. Calcite is more soluble if carbon dioxide is increased, acidity is increased, oxygen and organic matter are increased, temperature is decreased, pressure is increased, concentration of salts is increased, rate of flow is increased, and degree of mixing is increased.

The amount of carbon dioxide (CO_2) in solution is probably the single most important factor affecting solution because carbon dioxide combines with water to produce carbonic acid (H_2CO_3). The air, which normally has a pressure of 1 atmosphere, has a partial pressure of only 0.0003 atmosphere of CO_2. Rain water in equilibrium with air can dissolve very little calcite. Water containing oxygen and decaying organic material, however, can possess 0.1 atmosphere of CO_2 (over 300 times more CO_2 than normal rain water) and is able to dissolve a lot of calcite. It is possible to make undersaturated solutions simply by mixing two types of water having different pressures of CO_2, different salinities, or different temperatures. Undersaturation occurs in the case of CO_2 because a non-linear relationship exists between the partial pressure of CO_2 and the solubility of calcite.

Limestone Cave Formations

The formations which hang from the ceiling of a cave are stalactites; those built up above the floor of a cave are stalagmites; whereas those sheet-like, layered deposits on the walls or floors are flowstone. A column forms by the joining of a stalactite and a stalagmite. Together these cave formations are known as speleothems.

The origin and age of speleothems is a controversial subject. A popular theory for the origin of caves involves two stages. The first stage was when the cavity was filled with water, and solution of limestone occurred. The second stage was when the cavity or cavern *was filled with air, and deposition of speleothems began from solutions depositing calcium carbonate. A less popular theory is that there was only one stage in cave formation with solution occurring in the water-filled part of the chamber concurrently with speleothem deposition in the air-filled spaces.*

Radiocarbon (C-14) dating of speleothems has been used by some scientists to support the great age of cave formations. However, attempts to date carbonate minerals directly generally give deceptively old ages because carbon from limestone with infinite radiocarbon age (carbon out of equilibrium with atmospheric carbon) has been incorporated in minerals with atmospheric carbon.

Coalbrookdale by Night -Oil painting by Philippe Jacques de Loutherbourg 1801

Coalbrookdale is a side valley of the Ironbridge Gorge in the borough of Telford and Wrekin and ceremonial county of Shropshire, England, containing a settlement of great significance in the history of ferrous metallurgy. It is in the ancient manor and ecclesiastical parish of Madeley and said to be one of the birthplaces of the Industrial Revolution. It is home to the Ironbridge Institute, a partnership between the University of Birmingham and the Ironbridge Gorge Museum Trust.

The Iron Bridge, a notable icon of the Industrial Revolution, crosses the River Severn at the Ironbridge Gorge, by the village of Ironbridge,

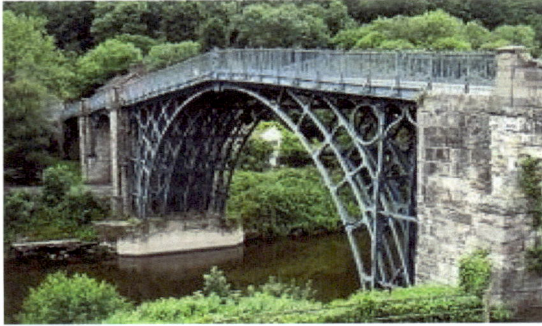

in Shropshire, England. It was the first arch bridge in the world to be made out of cast iron, a material which was previously far too expensive to use for large structures. However, a new blast furnace nearby lowered the cost and so encouraged local engineers and architects to solve a long-standing problem of a crossing over the river.

In 1709, Abraham Darby I rebuilt Coalbrookdale Furnace, and used coke as his fuel. His business was that of an iron founder, making cast iron pots and other goods, an activity in which he was particularly successful because of his patented foundry method, which enabled him to produce cheaper pots than his rivals. Coalbrookdale has been claimed as the home of the world's first coke-fired blast furnace and definitely the first in Europe to operate successfully for more than a few years.

Darby renewed his lease of the works in 1714, forming a new partnership with John Chamberlain and Thomas Baylies. They built a second furnace in about 1715.

The earliest iron artefacts were made by carving meteoritic iron. The earliest smelted iron object known is a sword from Tomb K at Alaca

Hüyük in Anatolia, Turkey and thought to have been smelted from charcoal and iron ore late in the 3rd millennium BC.

By the 8th century BC iron smelting became firmly established in the Middle East and the Eastern Mediterranean where it soon spread to Europe. Iron objects entered Europe some time before the technology to produce them. By the end of the fourteenth century, iron furnaces used in smelting were becoming larger with increased draft from large bellows being used to force air through the "charge" (mixture of raw materials). These larger furnaces first freed the molten iron in its upper levels. This metallic iron then combined with higher amounts of carbon because of the heated combustion blast produced by the air forced up through the furnace. The product of these furnaces was pig iron, an alloy that melts at a lower temperature than steel or even wrought iron. Pig iron was then further processed to make steel.

The Blast Furnace

The purpose of a blast furnace is to chemically reduce and physically convert iron oxides into liquid iron called "hot metal". The blast furnace is a huge, steel stack lined with refractory brick, where iron ore, coke and limestone are dumped into the top, and preheated air is blown into the bottom. The raw materials require 6 to 8 hours to descend to the bottom of the furnace where they become the final product of liquid slag and liquid iron. These liquid products are drained from the furnace at regular intervals. The hot air that was blown into the bottom of the furnace ascends to the top in 6 to 8 seconds after going through numerous chemical reactions. Once a blast furnace is started it will continuously run for four to ten years with only short stops to perform planned maintenance.

Iron oxides can come to the blast furnace plant in the form of raw ore, pellets or sinter. This ore is either Hematite (Fe_2O_3) or Magnetite (Fe_3O_4) and the iron content ranges from 50% to 70%. This iron rich ore can be charged directly into a blast furnace without any further processing along with limestone and coke.

Chemical Reactions in the Blast Furnace

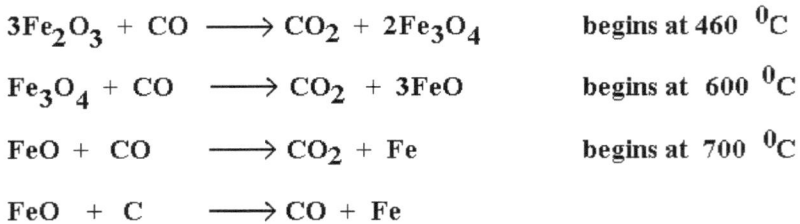

$$3Fe_2O_3 + CO \longrightarrow CO_2 + 2Fe_3O_4 \qquad \text{begins at 460 } ^0C$$

$$Fe_3O_4 + CO \longrightarrow CO_2 + 3FeO \qquad \text{begins at 600 } ^0C$$

$$FeO + CO \longrightarrow CO_2 + Fe \qquad \text{begins at 700 } ^0C$$

$$FeO + C \longrightarrow CO + Fe$$

The iron oxides go through these purifying reactions, they begin to soften, melt and finally trickle as liquid iron through the coke to the bottom of the furnace.

The coke descends to the bottom of the furnace to the level where the preheated air or hot blast enters the blast furnace. The coke is ignited by this hot blast and immediately reacts to generate heat as follows:

$$C + O_2 = CO_2 + Heat$$

Since the reaction takes place in the presence of excess carbon at a high temperature the carbon dioxide is reduced to carbon monoxide:

$$CO_2 + C = 2CO$$

The product of this reaction, carbon monoxide, is necessary to reduce the iron ore as seen in the previous iron oxide reactions. The limestone descends in the blast furnace and remains a solid while going through its first reaction:

$$CaCO_3 = CaO + CO_2$$

This reaction requires energy and starts at about 1600°F. The CaO formed from this reaction is used to remove sulfur from the iron which is necessary before the hot metal becomes steel. This sulfur removing reaction is:

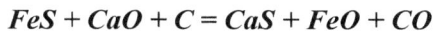

$$FeS + CaO + C = CaS + FeO + CO$$

The calcium sulphide becomes part of the slag. The slag is also formed from any remaining Silica (SiO_2), Alumina (Al_2O_3), Magnesia (MgO) or Calcia (CaO) that entered with the iron ore, pellets, sinter or coke. The liquid slag then trickles through the coke bed to the bottom of the furnace where it floats on top of the liquid iron since it is less dense.

Another product of the iron making process, in addition to molten iron and slag, is hot dirty gases. These gases exit the top of the blast furnace and proceed through gas cleaning equipment where particulate matter is removed from the gas and the gas is cooled. This gas has a considerable energy value so it is burned as a fuel in the "hot blast stoves" which are used to preheat the air entering the blast furnace to become "hot blast" or to generate steam which turns a turbo blower.

In summary, the blast furnace is a counter-current reactor where solids descend and gases ascend. In this reactor there are numerous chemical and physical reactions that produce the desired final product which is hot metal. A typical hot metal will contain the following mix of elements:

Iron (Fe)	*93.5 - 95.0%*
Silicon (Si)	*0.30 - 0.90%*
Sulfur (S)	*0.025 - 0.050%*
Manganese (Mn)	*0.55 - 0.75%*
Phosphorus (P)	*0.03 - 0.09%*
Titanium (Ti)	*0.02 - 0.06%*
Carbon (C)	*4.1 - 4.4%*

In 1996, China manufactured just over 100 Mt of steel and became the world's largest steel producer. In 2008 it produced in excess of 130 Mt of steel. Greenhouse gas emissions in the steel sector are primarily the result of burning fossil fuels during the production of iron and steel. Reduction of the iron ore is the largest energy-consuming process in the production of primary steel. In 1994, this process was responsible for over 45% of the CO_2 emissions from US integrated *steelmaking. Carbon dioxide emissions from steel production accounted for 9% of total carbon dioxide emissions in China in 1995. (Mt = million tonne)*

SCHEMATIC OF A BLAST FURNACE

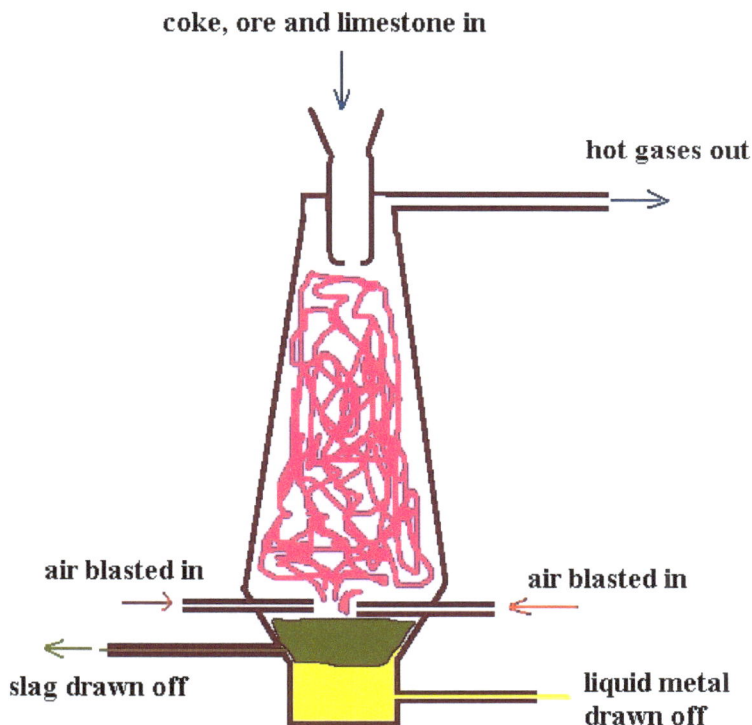

coke, ore and limestone in

hot gases out

air blasted in

air blasted in

slag drawn off

liquid metal
drawn off

Carbon intensity declined steadily from 1980 to the early 1990s, and has recently begun to fall again after a short rise, reaching 1.03 tC/t steel in 1996. Steel-related carbon dioxide emissions closely mirror primary energy use, with China clearly dominating, followed by India, Brazil, and Mexico. Carbon dioxide emissions from steel production are responsible for 13% of total emissions in Brazil, 12% of total emissions in South Africa and in China, 8.5% of total emissions in India, and 6% of total emissions in Mexico. In 2008 the World mean for steel-related carbon dioxide emissions was around 8 % of all carbon dioxide emissions.

The Manufacture of Portland Cement

In 1824, Joseph Aspdin, a British stone mason, obtained a patent for a cement he produced in his kitchen. The inventor heated a mixture of finely ground limestone and clay in his kitchen stove and ground the mixture into a powder to create a hydraulic cement i.e one that hardens with the addition of water. Aspdin named the product portland cement because it resembled a stone quarried on the Isle of Portland off the British Coast. With this invention, Aspdin laid the foundation for today's portland cement industry.

Portland cement clinker was first produced commercially in 1842 in a modified form of the traditional static lime kiln. The basic egg-cup shaped lime kiln was provided with a conical or beehive shaped extension to increase draught and thus obtain the higher temperature needed to make cement clinker. For nearly half a century, this design, and minor modifications, remained the only method of manufacture. The kiln was restricted in size by the strength of the chunks of raw mix, for if the charge in the kiln collapsed under its own weight, the kiln would be extinguished. For this reason, beehive kilns never made more than 30 tonnes of clinker per batch. A batch took one week to turn around i.e a day to fill the kiln, three days to burn off, two days to cool, and a day to unload. Thus, a kiln was able to produce up to

1500 tonnes per year. A kiln is basically an industrial oven, and although the term is generic, several quite distinctive designs have been used over the years.

In order to save money on fuel, a kiln was required that could run almost continuously, whilst the raw material was somehow fed through it. It was this scenario that lead to the development of the 'Chamber' kiln in the late 1850s. From 1885, trials began on the development of the rotary kiln, which today accounts for more than 95% of world production. The earliest successful rotary kilns were developed in Pennsylvania around 1890, and were about 1.5 m in diameter and 15 m in length. Such a kiln made about 20 tonnes of clinker per day. The fuel, initially, was oil, which was readily available in Pennsylvania at the time. It was particularly easy to get a good flame with this fuel. Within the next 10 years, the technique of firing by blowing in pulverized coal was developed, allowing the use of the cheapest available fuel.

Schematic of a cement Rotary Kiln

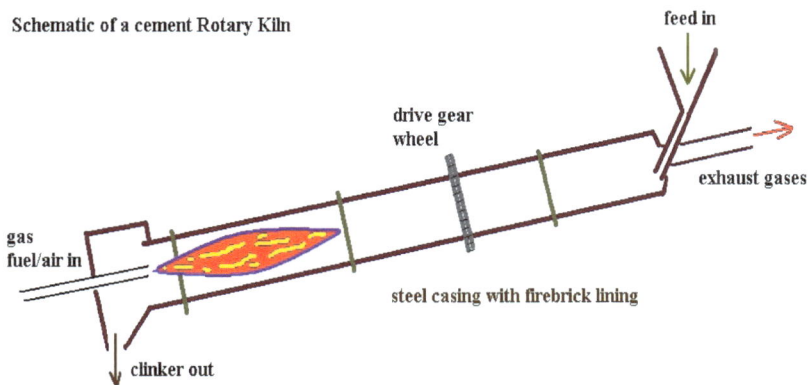

feed in

drive gear wheel

exhaust gases

gas fuel/air in

steel casing with firebrick lining

clinker out

Cement Kiln Emissions

Carbon dioxide: During the clinker burning process CO2 is emitted. CO_2 accounts for the main share of these gases. CO_2 emissions are both raw material-related and energy-related. Raw material-related emissions are produced during limestone decarbonation ($CaCO_3$) and account for about 60 % of total CO_2 emissions. Dust: To manufacture 1 tonne of Portland cement, about 1.5 to 1.7 tonne raw materials, 0.1 tonne coal and 1 tonne clinker (besides other cement constituents and sulfate agents) must be ground to dust fineness during production. In this process, the steps of raw material processing, fuel preparation, clinker burning and cement grinding constitute major emission sources for particulate components. While particulate emissions of up to 3,000 mg/m^3 were measured leaving the stack of cement rotary kiln plants as recently as in the 1950s, legal limits are typically 30 mg/m^3 today, and much lower levels are achievable.

Nitrogen oxides (NO_x): The clinker burning process is a high-temperature process resulting in the formation of nitrogen oxides (NO_x). The amount formed is directly related to the main flame temperature (typically 1850-2000 °C). Nitrogen monoxide (NO) accounts for about 95 % and nitrogen dioxide (NO_2) for about 5 % of this compound present in the exhaust gas of rotary kiln plants. As most of the NO is converted to NO_2 in the atmosphere, emissions are given as NO_2 per cubic metre exhaust gas.

Sulfur dioxide (SO_2): Sulfur is input into the clinker burning process via raw materials and fuels. Depending on their origin, the raw

115

materials may contain sulfur bound as sulfide or sulfate. Higher SO_2 emissions by rotary kiln systems in the cement industry are often attributable to the sulfides contained in the raw material, which become oxidised to form SO_2 at the temperatures between 370 °C and 420 °C prevailing in the kiln preheater. Most of the sulfides are pyrite or marcasite contained in the raw materials. SO_2 emission concentrations can total up to 1.2 g/m^3 depending on the site location.

Carbon monoxide (CO) and total carbon: The exhaust gas concentrations of CO and organically bound carbon are a yardstick for the burn-out rate of the fuels utilised in energy conversion plants, such as power stations. By contrast, the clinker burning process is a material conversion process that must always be operated with excess air for reasons of clinker quality. In concert with long residence times in the high-temperature range, this leads to complete fuel burn-up.

The emissions of CO and organically bound carbon during the clinker burning process are caused by the small quantities of organic constituents input via the natural raw materials (remnants of organisms and plants incorporated in the rock in the course of geological history). These are converted during kiln feed preheating and become oxidized to form CO and CO_2.

Cement production now accounts for around 7.5% of all carbon emissions worldwide.

To lean o'er stem and plough the sea
The breeze striking 'tween leach and luff
Fills my heart with joyful glee

The folding away of froth and wake
Glistening under clear moon and guiding star
My soul striving to venture afar.

It is interesting that Halley's Comet is considered a relatively large body compared to other comets. Although its nucleus is an odd shape, we can approximate it to a spherical shape of mean radius 5 km. It is thought to be comprised (as with other comets) of carbon dioxide, carbon monoxide, ammonia, carbon and carbon compounds, dust and rocks (silicates). Its total mass is about 10^{14} kilogram or 100 000 Mega tonne. If we estimate that its carbon dioxide/carbon monoxide content is about 55% by mass, this gives us 55 000 Mega tonne. How *does this compare with a normal amount of CO_2 in the Earth's atmosphere? What effect would it have if it were to collide with our planet? Well the mass of our atmosphere is around 5×10^{18} kg. or 5×10^{12} Mega tonne. So if it were evenly distributed it would increase our CO_2 by only one part in a hundred million... hardly a significant rise! The local damage however would be something different with a very large area completely devastated and perhaps a regional rise in CO_2 sufficient to poison plants and animals over an area of tens of thousands of square kilometres. If it crashed into the sea, the resulting tsunami would flow around the globe also causing tremendous damage and loss of life.*

Over the past one hundred million years, comet Halley has circled the sun in its long elliptical orbit one and a quarter million times. Each time it has shed off masses of material due to the solar wind, a constant stream of highly energised ions (charged particles). As it circles out to the outer reaches of our solar system it also collects debris such as more ice, dust and rocks. The planet Jupiter plays a significant role in the speed and orbit of comets. It may either capture a comet (as seen in 1994) or hurtle it back into space at a greater velocity. In 1992 the periodic comet Shoemaker Levy 9 made an extremely close passage of Jupiter. The stresses induced by the giant planet's gravity shattered the comet's nucleus, estimated to have been 7.5 km in diameter, into a dozen or more major fragments, the largest of which was about 4 km in diameter. Two years later, the returning fragmented comet crashed into Jupiter. Spectacular observations from both terrestrial observatories and the Hubble Space Telescope yielded vast amounts of data about the structure of comets and about Jupiter's atmosphere. Comet Halley was visible in 1910 and

118

*again in 1986. Its next perihelion passage will be in early 2062. I will
probably miss it!*

*Supposing then that Comet Halley was many times its current size
some one hundred million years ago; perhaps with a diameter of 100
Km. This would give it a volume of 1.6 million cubic kilometre and
with a mass of 4.7 x 10^{11} Mega tonne. Repeating our earlier
calculation of the effect of this
amount of material falling to
Earth, we get approximately 2
parts CO_2 per hundred parts
air, ie. 2%. Besides the greater
impact, this amount of CO_2
would certainly kill off larger
animals, increase the acidity of
natural waters in lakes and*

*oceans and cause a spiralling global warming. It is difficult to
speculate whether such large comets existed in earlier times. Current
evidence suggests that comets of this size tend to fragment near
Jupiter or near the sun. Analyses of Vega 1 and Vega 2 images of the
comet Halley nucleus led to a rotation period of 53.5±1 hours about
an axis approximately perpendicular to the long axis of the nucleus.
Images obtained by Giotto were also found to be consistent with the
Vega-derived rotation period. The approximate dimensions of the
nucleus are 16 km x 8 km x 8 km. (Giotto photograph   March 14th
1986)*

In 1996 the ROSAT satellite detected X-rays emanating from the Comet Hyakutake. It is believed that these were due to some interaction between the comet material and the solar wind.

Of course there has been much speculation over the event of "The End of the Dinosaurs" some 65 million years ago. Some of these theories include:

- *Exceptionally high cosmic radiation from a nearby supernova.*
- *Acid rain and gases from volcano activity*
- *Sudden high CO_2 levels from interaction of volcanic lava with limestone*
- *Super Ice Age*
- *A comet or asteroid struck the Earth*
- *Continental drift altered the climate.*
- *Some disease spread through the dinosaur population.*
- *The mammals appeared that began feeding on dinosaur eggs.*

Perhaps two or more of these factors hit the dinosaurs simultaneously ensuring their fate.

The Deccan Traps in India are one of the largest volcanic provinces in the world. It consists of more than 2,000 m of flat-lying basalt lava flows and covers an area of nearly 500,000 square km in west-central India. Estimates of the original area covered by the lava flows are as high as 1.5 million square km. The volume of basalt is estimated to be 512,000 cubic km. The Deccan Traps are flood basalts similar to the Columbia River basalts of the north-western United States. The

Deccan basalts may have played a role in the extinction of the dinosaurs. Most of the basalt was erupted between 65 and 60 Mya (million years ago). Gases released by the eruption (such as sulphur dioxide and sulphur trioxide) may have changed the global climate and lead to the demise of the dinosaurs. Firstly they would have had difficulty breathing in the poisonous air. Secondly, the dust and debris hurled into the atmosphere would have brought on a sudden cooling of the Earth's surface.

In 1979 Luis and Walter Alvarez studied the layer of sediment which was deposited between the Cretaceous and Tertiary Periods. They found extremely high concentrations of the element iridium. This element is several thousand times more abundant in meteoric dust than it is on the surface of the Earth.

They theorised dinosaur extinction was caused by an asteroid. In the 1980's the father-son team discovered a layer of Iridium in the Cretaceous-Tertiary boundary. Iridium is rare on Earth, but abundant in meteorites. The Alvarezs suggested that a huge asteroid or comet, several kilometres in diameter struck the Earth at that time. The result of such an impact would be an enormous explosion throwing dust clouds into the sky darkening the

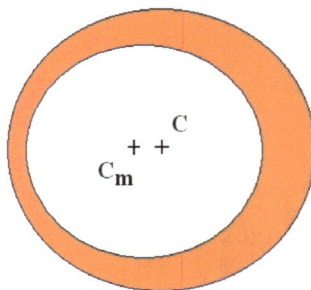

Crossection of Moon

C_m centre of mass

C geographic centre

planet. Massive forest fires, triggered by the hit, would add smoke to the sky. This would cool the planet causing the climatic changes observed- a sudden ice-age! A crater, now worn down and partly under the ocean, was found along the Mexican Yucatan Peninsula. Its appearance coincides nicely with the Cretaceous-Tertiary boundary. NASA scientists estimate that the asteroid that struck would have been about 16 km in diameter. The Crater is about 200 kilometres across.

Did the Earth once have more than one moon? Is it possible that a smaller moon existed inside the current moon's orbit? Was this the asteroid? As moons of planets nearly always lie in the same plane, we need to calculate the estimated radius of the current moon's orbit 100 million years ago and see if the mathematics allow for another smaller satellite. How did such a moon's orbit diminish? We know that the crust of our moon is thicker on the further side, causing it to rotate in such a way that only one side faces the Earth at all times. Is the moon an aggregation of other moons of the Earth at greater distances? Perhaps we will discover this to be true. Our moon is so large that astronomers prefer to describe our unique system as a double planet. So did our Little Lost Sister Moon crash to Earth or Moon?

Earth

Little Lost
Sister Moon

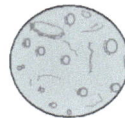

Moon

Is the diameter of the Earth constant? Has it been larger and/or smaller in the past? We know that at the Earth's core there are complex nuclear reactions going on. Do these meet some critical stage where evolutionary production of elements flips to a new state causing an expansion or contraction? A contraction of just two hundred metres would bring about a flooding of a third of the land surface. We know that the Earth's magnetic field has flipped over the eons and that continental plates are continually slipping over or under each other so the theory is not preposterous. We just need to look for evidence. Why did the dinosaurs grow so large? We do not see animals of such dimensions now!

Antarctica

Due to its massive size and major source of fresh water locked as ice, Antarctica seems to be a focus for climatologists and researchers investigating the history of the Earth's climate and factors such as the chemical composition of the atmosphere over millennia. It is regarded as a key indicator of global warming although there is still some disagreement about observed trends over the last five decades. Another perspective is the eyes of nations upon the likely huge resources in Antarctica such as fresh water, minerals, oil, gas and coal reserves seen to be dwindling in many parts of the world.

A catastrophic effect of global warming would descend if large amounts of water bound up in ice in Greenland and Antarctica melted and raised the seas. University of Colorado at Boulder researchers using data from a pair of NASA satellites orbiting Earth in tandem have determined that the Antarctic ice sheet, which harbours 90 percent of Earth's ice, has lost significant mass in recent years.

The team used measurements taken with the Gravity Recovery and Climate Experiment, or GRACE, to conclude the Antarctic ice sheet is losing up to 150 cubic kilometres of ice annually. By comparison, the city of Los Angeles uses about 4.5 cubic kilometre of fresh water annually.

This is the first study to indicate the total mass balance of the Antarctic ice sheet is in significant decline. At the measured rate of melting it would take about 6 years for the oceans to rise 25 mm or 72 years to rise 300 mm.

*The Intergovernmental Panel on Climate Change assessment, completed in 2001, predicted the Antarctic ice sheet would **gain** mass in the 21st century due to increased precipitation in a warming climate. Other studies signal a **reduction** in the continent's total ice mass, with the bulk of loss occurring in the West Antarctic ice sheet.*

But, as data from ice-core has shown, there can be a lag of anything between a hundred and a thousand years between effects. This presents the difficulty in making absolute predictions.

Further, researchers used GRACE data to calculate the total ice mass in Antarctica between April 2002 and August 2005. Launched in 2002 the two GRACE satellites circle the Earth 16 times per day at an altitude of 510 km, sensing subtle variations in Earth's mass and gravitational field. Separated by a constant 225 km, the satellites measure changes in Earth's gravity field caused by regional changes in the planet's mass, including such things as ice sheets, oceans and water stored in the soil and in underground aquifers.

Retreat of Glaciers on Antarctic Peninsular

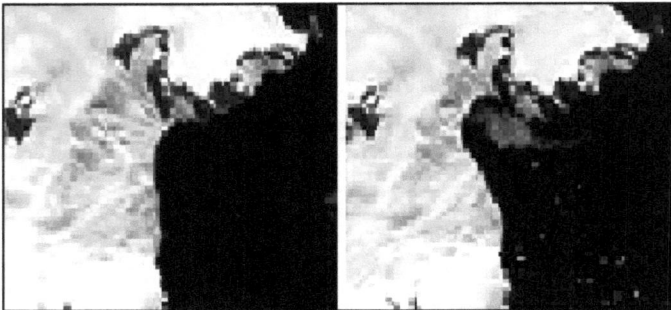

1981   2001

Subtle changes in gravity due to a pass over a portion of the Antarctic ice sheet, for example, pulls the lead satellite away from the trailing satellite. A sensitive ranging system allows researchers to measure the distance of the two satellites down to as small as 1 micron and calculate the ice mass in particular regions of the continent. These satellites enabled collection of data across the entire Antarctic. While the researchers were able to differentiate between the East Antarctic ice sheet and West Antarctic ice sheet with GRACE, smaller, subtler changes occurring in coastal areas and even on individual glaciers are better measured with instruments like radar and altimeters.

A study published in September 2004 concluded that glaciers on the Antarctic Peninsula sped up dramatically following the collapse of Larsen B ice shelf in 2002. Ice shelves on the peninsula, which has warmed by an average of 2.5 degrees Celsius in the past 50 years, have decreased by more than 14000 square km in the past three decades.

The thickness of the Antarctic ice averages 2 km for the entire Antarctic. Antarctica is twice the area of Australia and contains 70 percent of the Earth's fresh water resources. The ice sheet, covers about 98 percent of the continent. Floating ice shelves constitute about 11 percent of the continent.

The melting of the West Antarctic ice sheet alone, which is about one eighth in volume of the East Antarctic ice sheet, would raise global sea levels by around 5 to 6 metre, according to researchers from the British Antarctic Survey.

What is most important about this study is that it uses satellites to build a much more comprehensive picture of what is happening with

126

the ice. The problem is that the satellites were launched only in 2002 making it impossible for accurate comparison with earlier data.

Richard Alley, a Pennsylvania State University glaciologist who has studied the Antarctic ice sheet but was not involved in the new research, said more research is needed to determine if the shrinkage is a long-term trend, because the new report is based on just three years of data. But Alley called the study significant and surprising because a major international scientific panel predicted five years ago that the Antarctic ice sheet would gain mass this century as higher temperatures led to increased snowfall. (I have already commented on a definitive lag-time between climate' cause and effect')

Scientists cannot prove beyond doubt that human activities are causing global warming. However, humans are changing the atmosphere on a large enough scale that the possibility exists that we are changing the climate. A melting of the Antarctic ice would produce huge costs all over the world. Coastal lands are valuable. Large low lying areas would be lost.

The problem we have with the climate debate is that even the most complex mathematical models do not predict reality since even these models contain many simplifications and assumptions that may be incorrect. Nature continually makes the climate change without human participation. So even once a change has happened it is still impossible to figure out how much of the change was caused by humans and what proportion is due to natural effects.

There is a need for cheaper non-fossils ways to create energy. Once such technologies are developed they will both reduce greenhouse gas emissions by displacing fossil fuels for energy and provide the energy we need.

We need greater accuracy and precision for measuring climate. The satellites used in this latest report demonstrate how much technology can help to answer questions which otherwise form the basis of speculation.

We need to develop ways to engineer the climate so that if the climate ever starts going directions that will cause huge problems for the human race we will be able to intervene and push it in a different direction. For example, if we could deflect light away from the Antarctic and from Greenland and we could prevent and even reverse ice melts.

We see a similar story from a British study. About 212 of the 244 glaciers surrounding the peninsula, which stretches north from the southern polar continent toward South America, have retreated as temperatures have risen more than 2.5 degrees Celsius over the past fifty years

The glacial retreat puts Antarctic ice shelves and sheets at risk, states Alison Cook, a senior geographic data analyst with the British Antarctic Survey in Cambridge, England. Inland glaciers that flow from mountains into the ocean and keep continental ice sheets in place are retreating. This trend will continue if the climate continues to warm.

The Antarctic Peninsula is a mountainous region that extends from the 14 million-square-mile continent and ends 600 miles from the tip of Argentina. Its eastern side is flanked by the Larsen ice shelf, a floating ice mass that broke from the main continent in 2002. But can we be certain of a continued decline? We cannot.

But with the loss of ice at both poles we see a diminishing reflective power of the planet. This means more energy absorbed onto the land surface and the oceans in Polar Regions. Along with increased carbon dioxide emissions, the problem of a warming planet is further aggravated; the loss of such a major reflective surface like the polar ice caps could accelerate climate change. Antarctica may be shrinking as global warming encroaches, but its precious natural resources have led to increased interest in the seabed around it. The UK will submit a claim to the United Nations to extend its existing Antarctic territory by a million square kilometres. This is one of five territorial requests planned by Britain ahead of a May 2009 deadline and covers a vast area of the seabed around British Antarctica.

It is being described as the last big carve-up of maritime territory in history, with the object of acquiring potential areas for the mining of new reserves of oil and gas.

Around 7 million sq km, an area equivalent to the size of Australia could be divided between nations around the world. The attempt to extend British sovereignty in Antarctica could spark disputes with South American nations such as Argentina and Chile, who are likely to make overlapping claims in the region. As we have seen, a war has already been fought over disputed territory in the South Atlantic.

The measure conflicts with the spirit of the 1959 Antarctic Treaty, to which Britain is a signatory; it prevents the exploitation of oil, gas and minerals, other than for scientific research. Reminds me of the Japanese and their whaling program! But, with dwindling energy resources, such niceties have been relegated to history so it appears. Australia, Britain, Brazil, France and Russia are among those to have made claims, with Moscow planting a Russian flag near the North Pole in Arctic waters more than four kilometres deep. I suppose I could be an arch-cynic and lament that with no ice, exploration and mining on the Antarctic continent itself might be substantially made easier!

courtesy David Shultz

130

Again, I mention that in the prequel "Nuclear Islam" there is much information on the topic of nuclear power, specifically by conventional fission reactors. Whereas the author is firmly against the use of uranium, plutonium and other related materials for the production of electricity, it is apparent that many governments around the world have implemented nuclear power as an important source of their electrical power. The strongest argument against nuclear fission is the lack of a world coordinated solution to the hundreds of thousands of tonnes of lethal radioactive waste with the potential for health concerns to humans lasting thousands of years ahead. Individual nations have their own backyard disposal programs and frankly they are, in the main, doing a bad job of it. There is much written material available to suggest that Russia and its former territories of the USSR together with many other countries have not disposed of their waste safely and further, that there remain serious problems with aging reactors. Even the security issues are questionable with a very real threat of material falling to the hands of political extremist groups and terrorists. One smallish reactor in the UK that was decommissioned in the late 1990's is still undergoing site clean-up to the cost of some 20 to 30 million pounds. The green light day will be around 2025, in excess of twenty five years after closure. Other eastern European countries have not been so fussy or careful with evidence in one case of such minimal action being no more than a barbed wire fence, a patrolling guard with a tired Alsatian dog!

The proponents of nuclear power now use the "clean and green" slogan to promote this option. Aside the highly polluting radioactive waste is the secondary pollution of cheap and abundant electricity. Large populations and cheap electricity has already led to an explosion in industry, particularly in China and India, producing secondary emissions and pollutants to the environment. The momentum and attack on a shaky and overburdened atmosphere, soil, stream, river, lake and ocean by chemicals is to be the hallmark of the twenty first century despite the warnings and data from the earlier two centuries. Whereas the West has tightened considerably its environmental laws with agencies scouring the country for offenders and dealing out large penalties, the new third world with its disproportionately large populations cannot enact regulations fast enough to cope with the tsunami wave of industrialisation in their respective countries. This is the nub and crux of the problem to be born by the peoples of the whole world over the next seventy years. The predictions already made are dire. I suppose one could simplify this further and merely state that the true nature of our woes is population numbers generally coupled with the desire to improve the quality of life judged on the basis of material possessions. The average Jo Blow in Africa, South America, India and China wants all the goodies and trappings seen in the West at the beginning of the current century. They want to own a car, a house or apartment, electrical appliances deemed to make life more comfortable lumped together with all those silly fashionable designer labels that separate the haves from the have-nots.

Rolex, Omega, Guess, Armani, Ermenegelda Zegna, etc. Soon they also learn that for their children, to give them the best chance in life, they need private education and an international university. The very worst elements of society in European nations of the nineteenth century are emulated and spawn to create new divisions of class. Thus all the revolutions of the twentieth century are likely to be repeated at some later date in the twenty first century due to the emergence of elite classes and hatred from below. Global warming and pollution will likely unsettle societies and bring on confrontation between and within nations. Perhaps this is part of some anthropological law that has escaped the attentions of PhD thesis writing students of sociology, but inescapable by humans in general.

Heat energy is associated with nuclear power plants. A lot is wasted to the environment. The inefficiencies of industry also contribute to vast amounts of heat energy escaping into the environment. Even with electricity as the driver of our factories, they still need to make products. Products themselves create waste and emissions. This is not a criticism of our manufacturing industries. No, just to comment on the problem when it is multiplied a trillion fold across the planet. The sheer weight of industrialisation is the problem. So coupled with greenhouse gas emissions from hydrocarbon fuels (whether originating from petroleum oil or bio fuels is irrelevant), the horrendous increase in the need for energy and electrical power to drive the needs of humans remains the problem and cause of our future woes.

How can we reduce the demands and these perceived needs of humans? Will people suddenly and in a revolutionary way strive for a simpler and environmentally friendlier life style? How can the intellectual elements of the West suggest to the developing nations "Don't be like us, take greater care of your environment and settle for the simple life"? It's just not going to happen. In any case, the giant corporations and logo promoters want as big a slice of the new world as they can get; even vying to take a certain amount of control. Capitalism is linked to imperialism. Absolute power does corrupt. Governments with some standards, morals and sensibilities find it difficult to fend off these powerful Barons of the global giants of the commercial world.

Let us put these arguments aside and suggest that we do indeed need a lot more electricity and we wish to avoid as much as possible the need to burn fossil fuels such as coal and oil in order to reduce the amount of carbon dioxide reaching the atmosphere. Well I have already discussed other alternatives which are slowly taking hold around the world viz. wind, gravitational potential (hydroelectric and tidal electric), solar and geothermal. Where societies and governments still cannot meet their needs for energy and electrical power, the other option is nuclear fusion. The joke, which I repeat here, is that when scientists discuss nuclear fusion, they always retort that it is just twenty five years away and will be our salvation. They have been saying this since the early nineteen fifties however.

Despite this cynical remark, the author does have faith in that the 'nut' will eventually be cracked and humankind will suddenly benefit from an unlimited supply of unprecedented amounts of energy from this source. Just 1 kilogram of Deuterium and three kilogram of Lithium can theoretically produce an equivalent amount of energy as about 25000 tonne of coal. The details of fusion reactions are given in Appendix B but basically, by gluing light atoms together rather than splitting heavy atoms such as Uranium 238, huge amounts of energy can be created. The United States and the UK have been researching this possibility since the mid 1950's. The "Zeta Coil" was an early attempt at creating a plasma (superheated material to some 100 million degrees) to create fusion reactions plus energy. Unfortunately none of the machines created an excess of output energy over the input energy required to drive the machine. This has been the bane of all the research fusion machines to date. To sustain fusion, the temperature, density and confinement time must be simultaneously achieved. The combination or product of these three variables must surpass a critical quantity known as the Lawson Criterion. The positively charged plasma ions must be superheated until their energies exceed the electrostatic repulsive forces between them. High magnetic fields are applied to contain the plasma in a circular stream away from the relatively cooler vessel walls. The probable most significant factor in the way of the break-even point for all earlier machines was their size.

ITER (International Thermonuclear Experimental Reactor) is to be the first of the next generation of fusion plants and is currently under construction at Cadarache, in southern France. It is proposed to build a machine capable of 500 Mega Watt of output power in a joint venture between the US, UK, Russia, China, India, Japan, South Korea and the European Union of Nations. The budget is estimated to be around ten billion US dollars and it is expected to be operational by 2015. The crucial element of the machine is the tokamak, a vessel that suspends the plasma with magnetic fields and creating a similar effect on the plasma to that of the Sun's gravitational field (essential for its fusion of isotopes of hydrogen). This has already been achieved on a smaller version of the machine near Oxford in the UK and at Princeton in the USA.

When will we see the first commercial venture of a fusion reactor pumping electricity into a nation's power grid? Why in just another twenty five years of course! Joking aside, if this time the output is far in excess of the input energy, a new age will dawn. Fusion reactors will not bring such waste as conventional fission reactors however they are not without waste. The very components of the reactor themselves do become radioactive with a half life in the hundreds of years rather than thousands of years. The amount of this material will be just a fraction of that produced from fission reactors. Sadly, due to the lead time into widely spread commercialisation of fusion reactors, in the short term we are likely to see a steep increase in the building of new conventional nuclear power plants across the world after a brief lull which is soon to end. All these predictions are made in "Nuclear Islam" with graphs and accompanying data.

It has been said that a plasma fusion reactor cannot possibly produce a nuclear explosion or melt-down catastrophe such as experienced in Chernobyl, Ukraine. But then future fusion reactors on a commercial size have yet to be built and despite assurances from the scientists, we cannot claim to be 100% certain of that stance. Given enough plasma and high fusion rates, can one say with confidence that the device will not run away as with the thermonuclear hydrogen bomb? We all know the destructive power of such a bomb.

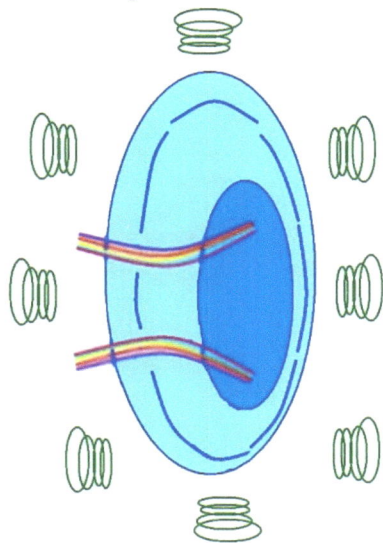

As aforementioned I give some further detail to the nature of fusion reactions in Appendix B for the interested reader. It is not comprehensive as there are hundreds of possible combinations of fusion reactions of the light elements. Suffice to say that this was the evolutionary process in the galaxy that led to the creation of all the heavier elements we know and find today distributed on our planet.

Seven   *Third World Abuse*

The nations of the West are mostly moving towards a greener and cleaner environment. As long ago as the late 1950's, Britain introduced the "Clean Air Act" as a response to the killer smog earlier that decade. The new act place responsibilities on companies and coal fired power plants to dramatically reduce their emissions of carbon dioxide, oxides of sulphur and other harmful effluents to the atmosphere. As people became more concerned generally with their nation's environment we have seen major shifts of heavy industries such as steel making, ship building and similar away from the West to the East. Basically we have replaced the smogs of a post-war West with the smart-arse smugs of the 21st Century. The dirty businesses have been moved off-shore to the poorer nations where labour is cheaper and government regulations and restrictions have not caught up. The price of a cleaner West has been to the detriment and environmental degradation of a poorer East. From personal experience I have noted the canals and streams of Indonesia, the land degradation in India and the cheese-like atmosphere of China. Heavy metals, organic solvents, plastic and paper waste, high levels of CO_2 and SO_2 are prevalent in all these countries. The conglomerates of Japan, Germany, USA, UK, France and a host of other richer nations are happy to see their products manufactured in these poorer countries to make bigger profits with fewer overheads. Cars, motorbikes, trucks, petrochemicals, fertilizers, dyes, fabrics, paints and coatings, shoes, tires, designer label clothes all made in the sweat-shops of the East, South East Asia, Africa and South American nations. Child labour is still a problem that UNICEF struggles to *combat (not to mention Child Soldiers). One of the solutions to a depletion of oil supplies and a consequent rise in the price of oil is the production of bio fuels. We now see millions of acres of forests disappearing in the same nations to produce crops for the production of alcohol to fuel automobiles in the West. Often these areas farmed for fuel are in countries where food production is minimal and people are going hungry for the want of it. Timber resources are also being cut down illegally in Kalimantan, Sumatra and other parts of Indonesia to satisfy an export market to Japan and other rich nations. The story is repeated in Western African nations, Central and South American nations and other parts of South East Asia. Forests and Wetlands (bogs and swamp habitats) are being cleared for palm oil and other crops. The rate of deforestation is profound and species of tree, plant, grass, shrub as well as animals, birds and reptiles is at an unprecedented and alarming rate. The guilty are often international corporations with their headquarters in Tokyo, New York, London and Berlin. OK, I know that North America and Europe cleared many of its swamplands and bogs to reclaim for pasture and crop growing. Much of this was accomplished hundreds of years ago and we have the same predicament of telling others not to do what we have already done. It is a dilemma not easily solved. How does one view the morality of the West when we become aware of these entire goings on? Take the case of the national parks in Sumatra. The Indonesian government set aside certain lands for development, mainly into palm plantations for palm oil. They defined quite clearly the borders of national park never to be intruded upon. So what happens? Small entrepreneurs eat away at the national park with small plots ranging from half a hectare up to ten hectare and clear fell to plant palm. This continues for a few years with local police and government agencies taking bribes to turn a blind eye. The final stage is an amalgamation of these small blocks to large scale plantations. And who do you think are the owners of these? Why the very same owners of the legitimate plantations outside the national park. One might suspect a conspiracy from the outset where national parks are not seen by the developers as worth while. Meanwhile the very heritage of Indonesia is being whittled away by greedy corporations and corrupt officials. But this situation is not unique to Indonesia (cf. Brazil's Amazon jungles). It is common throughout the third world and often the guiltiest are to be found in the first world. So much for the policies of the smug and self-deluded Western nations. It is all in vain as the surface of this planet is finite and whatever goes on environmentally at the other side of the world will eventually affect us all.*

But all is not entirely lost. Time may not be on our side but there are still things we can do and there are many people making a positive contribution. Education is probably the most important ingredient to save our future. An awareness of those companies that have a clean and tidy front, slick marketing and carefully worded rhetoric to fool the majority of us but have a murky backyard and disturbing record of abuse of the environment in the third world. Exposure is the best tack followed by heavy penalties. Of course sacrifices must be made and unfortunately it is in the nature of most of us not to give up our creature comforts or step down a little on the ladder of wealth. We are not prepared to reduce our energy use or resource extravagance but these things we must compel ourselves to do if we are to have a fighting chance. America has not signed the Kyoto protocol and was severely embarrassed at the Bali Conference at end of 2007. Their federal government seems powerless to make political inroads to changing practices that are proven threats to the environment. (There still exists more than a handful of Senators that do not believe in climate change and global warming due to human activities). But we now see the American people themselves taking up the sword and challenging this view (hopefully, at the same time, putting down the gun!). Al Gore has made a lasting impression on America and it is small town local governments and state governments that are now beginning to make the difference in America. This is a heartening picture to see a resilient people defy a conservative and immovable central authority of procrastinators. There has been general agreement at the Paris 2015 conference, but nations must make their own decisions and destinies regarding reducing carbon emissions. TL has suggested that exporting nations of fossil fuels should include 50% of these in their own national emissions calculations. This would not go down well for countries such as Kuwait and Australia where they have large exports of fossil fuels but small national populations!

Interesting that one of the largest companies in the world to produce solar panels is in Wuxi, Southern China, not too far from Shanghai. The sad thing is that most of their product goes overseas with just a trickle to the local market. But a trickle will grow in time to a river and finally a deluge. I noticed in Australia that the giant British oil company BP is a major player in solar panel production and sales. I have to admit I have mixed feelings about that as they have a strong vested interest in oil and its derivatives. Also, can they keep the price of solar panels high? But it does not surprise me. Energy is their name and energy is their game!

The wars in Iraq/Syria and threats to Iran can and have been viewed as a bold plan by certain countries to get back control of Arab oil fields in a world of dwindling volumes of that resource. If this was the case, it presents a scary picture of our future as energy resources do get down to critical levels. Leaders of nations should take a second look at the ideas presented in the chapter entitled "Futureland" in the prequel Nuclear Islam by yours truly. A lot of careful planning and preparation is necessary now for each nation to take responsibility for its own well being in the coming crises of shortages of ... EVERYTHING!

The rise of religious extremism is also discussed and its implications as another factor in the decades to come.

There is sufficient evidence that the US secret service agencies have played a role in destabilising governments of other countries by what would now be labelled as "terrorist means". These include economic actions, political assassinations... to supply of arms, money and military servicemen. Where a country was seen not to fit in with American Foreign Policy, a clandestine response of considerable magnitude was seen as the best thing to do in the interests of America. Did the "interests of America" go beyond political differences in the sense of a difficult high road for American investment and opportunity? Some analysts would give more than a strong nod to this suggesting that American foreign policy at times could only be defined as being self-interest to the point of imperialistic. The worst of this occurred in the latter half of the 20th Century. Now, the American people tend to question more "why is the government doing this?" Interference in the third world still goes on. Some of it is justified. The petty dictators of the Balkans and some African nations deserve to be thrown out and incarcerated for their evil doings. No-one sheds a tear for the Idi Amins and Sadam Husseins of the world but we cannot allow ourselves to posture as the angels of good and mercy whilst at the same time, corporations under our national banner are poisoning the land of the third world by profligate practices now outlawed at home. We all know the long term effects of the pesticide DDT... this agent has now returned to parts of the third world for crop dusting and protection. How did we get to such a situation?

Bangkok, Jakarta, Bombay and Manila have become notorious cities where Caucasian and Japanese men tour for sex holidays, sometimes leaving behind their unwanted offspring or else bringing home disease to their wives or partners. Paedophiles haunt the alleys of these cities alongside child molesters from the West. These are small in numbers but never-the-less take part in the further abuse of the peoples of the poor nations. In Russia, the Ukraine and some European countries, home grown mafia groups control vast empires in prostitution explicitly for Western gentlemen (?). Some women do manage to escape and enjoy a more normal life. (See details in "From Russia with Love" and unpleasant solutions proffered in "Return to Animalia" each by the author)

The testing of nuclear devices in the Pacific and Australia far from the shores of the participating nations might also be seen as an abrogation of the laws or promises to protect those cultures in close proximity to the testing grounds. These acts of arrogance by either an unthinking or uncaring nation over other cultures are unforgivable.

The sale of pesticides and seeds for the growing of food to poorer nations has not always been to the benefit of those nations. The loss of their original seed varieties and reliance on chemical companies has

143

not always brought about an improvement to the economy of the country but in some cases brought about additional debt.

The money lenders to the third world have also brought wealth to themselves and further poverty to those nations that find themselves unable to pay back the interest, making no dent in the borrowed amount at all. But then we can alleviate the situation just a little by getting these poor nations to further sign up for "lucrative" armament deals. Some debt by the third world to the first world has indeed been written off. But how many times have there been "conditions"? Slush, fudge and mud!!

This list of abuse to peoples of the third world goes on and on and it is no wonder that extremist political groups spring up that no longer wish to part of the West's "Big Plan". And what then is our response to that? Slag them off in the media, label them terrorists (instead of National Awareness Revolutionaries) and eventually crush them by war into final submission. After all it is we that are the saviours of the world and only we know what is good for them. And we can pat ourselves on the back spouting self delusive platitudes as the true champions of democracy and all that is righteous and correct. "The guardians of the only true path to enlightenment and prosperity".... and to hell with the rest or any other point of view!

The most fearful ingredient now inflicted upon the poorer nations is economic recession where so much idealism will fly out the door under the banner of 'we just can't afford it right now, it'll have to go on the back burner'. The richer nations will look to their own woes and abandon their poorer neighbours by putting them out of their minds or relegating any newsworthy plight to a footnote after the 'sports news' in the printed media; alas and alack!

Even within the richer nations, in extreme economic downturn the ideals of a cleaner world may also drift from focus exacerbating carbon emissions instead of damming up the wall. The outcome and consequences of this tact I cannot bear to think about! We must adhere to the little that has been promised and not shy away from those targets set at world environmental conferences in recent years else the worse case scenarios spelled out will be accelerated!

Finally, with the combined efforts of Russia and the West, civilisations in countries such as Afghanistan, Iraq, Libya and Syria have been systematically reduced to rubble. Millions have died and tens of millions have become refugees seeking safer havens in new countries. The stresses and strains on the whole global society unfathomable along with human misery on an unprecedented scale! The capitalistic venturism (my word) of companies producing weapons and ammunition only exasperate the situation. A Kalashnikov manufactured in China now sells for a mere fraction of the price 50 years ago. A Glock pistol manufactured in America can be had for just a couple of hundred dollars! Thousands are dying each week in these perpetuated wars, but the sophisticated media sensationally sweeps all this knowledge away after a handful of retaliatory deaths of us Whiteys in the West resulting in a sudden knee jerk response out of all proportion. We shall reap what we sow!

... and the last Trump was heard to blast...

(from a very fat man in America.)

The emerging economic giants we know to be China and India. They are giants due to their incredible populations and their current surge to dominate the commercial world. They already contain 38% of the world's population between them. It is thought that the population of India is likely to surpass that of China before 2020. But this is not the essence of this chapter. This chapter is in response to the greatest competition ever: "to find a way of removing ten billion tonne of carbon from the atmosphere each year" This competition and prize has been set up by Sir Richard Branson of the famed Virgin Air Corporation. Tom Law as an author represents the publishing company Longership Publishing Australia in this competition and this chapter outlines his solution.

Sequestering of Carbon.

We have already seen that the laying down of coal and petroleum oil took place over a time period of 400 million years. Most of this hydrocarbon we will find exhausted well before the end of the current *century. This equates to a time span of just three hundred years in which we will have burned the total fuel laid down in 400 million years. It is not necessary to have earned a doctorate in Mathematics to see that there are likely to be dire consequences to the Earth as a result. It has been argued that Mr Gore's book "An Inconvenient Truth" has scientific flaws in it and I tend to agree with this analysis. But they are subtleties and do not hold great sway on his fundamental truth! His examples of extremes in weather at various parts of the globe do not necessarily point to global warming and climate change as a response to human activity. However there is now an overwhelming amount of evidence to suggest that there is a radical departure from gradual change in climatic conditions, particularly CO_2 levels and mean temperature rise, indicating that we are in for a rough ride over the next few decades if we do not dramatically change our bad habits of excessive emissions of greenhouse gases into the environment, the atmosphere in particular.*

It could simply be argued that there are too many people on the surface of the planet and that by removing nineteen twentieths of the current population would immediately solve the problem. But this would mean the sudden extermination of some 6.08 billion persons with a remainder of 320 million to carry on. This is probably not the solution that Sir Richard had in mind so I will outline a more humane solution.

To do battle against these changes I outline some suggestions and provide some calculations which hopefully meet Sir Richard Branson's requisite target:

1. Pooh Bear to be Relegated to Ancient History or the Revenge of the Honey Bee.

Honey is predominantly sugar with a smallish percentage of water. We may simplify the formula of honey to its empirical formula CH_2O which has an empirical mass of: $12 + 2 + 16 = 40$ relative atomic mass units. We can see also that carbon takes up 12/40 parts or 30% by weight. Taking into account a certain amount of water, this equates roughly to 25% of carbon by weight. Now the honey bee is a friend of nature and a considerably hard worker. Consequently she has a very short life of around two months on average. She assists with the pollination of flowering plants which of course includes trees. Thus my first suggestion is that the human race desist from consuming honey but rather permit all honey to go into safe storage, possibly in a secret location somewhere in central Australia or better still Antarctica where my next friend (the ant) cannot exist to consume it. Bees wax contains a massive 85% carbon by weight ratio. Of course we must increase the number of hives throughout the world and elevate apiarists to one of the more highly paid professions. Universities worldwide must expand their faculties to include the 'Faculty of Apiary'. The whole technology must be revolutionised.

Considering the Earths land area and suitable climes for the honey bee, available vegetation and nearby source of fresh water, I calculate that w million tonne of honey, bees wax and bee acid can be stored per annum in the secret location. This amounts to x million tonnes of carbon stored. Of course this honey must be paid for and all governments of the world must contribute to the storage as a form of taxation.

2.   Troops of Ants

*It has been known to chemists for thousands of years that the body of an ant contains a certain percentage of the organic acid Formic Acid, whose formula is: HCOOH or more simply: CH_2O_2 The empirical mass of formic acid then is: 12 + 2 + 32 = 46. This equates to carbon being 12/46 parts by weight or 26%. Again, allowing for a tad of water, we may say that concentrated formic acid contains 25% by weight of carbon. Thus if governments encourage ant farms for the production of formic acid (simply done by distillation of the bodies of the ants) we could collect millions of tonne of formic acid per year. Again this could be stored in a similar secret location for eternity (alongside our spent plutonium and uranium). Incidentally, there was a huge scandal in China just recently where people were encouraged to invest in ant farms with the promise of big returns. People invested in the ants but the entrepreneurs disappeared with the investors stuck with their ants and no way of producing an income. It was a giant fraud. However in my scheme, governments will guarantee the buying of the formic acid at some favourable royalty as part of their commitment to the carbon tax.. Again I have made a similar calculation and arrived at the figure of y million tonnes per annum of formic acid production, ensuring the storage of z million tonne of carbon. Of course more research can be done to discover the most favourable strain of ant for maximum acid production. Different types of ant might be more suited to different environments and habitats. Whole **islands** might be devoted to formic acid production.*

3.   Titanic Ocean Skimmers

I have read articles on the idea of seeding the oceans of the world with various salts, iron in particular, to stimulate the growth of *plankton and other living creatures that dwell in the top metre or so of the ocean surface. The suggestion is that the additional bloom will eventually die and the carbon containing carcases fall to the bottom of the ocean thus storing more carbon. The flaw in this might be that the carbon never reaches the bottom of the ocean but instead enters the food chain and eventually finds its way back into the atmosphere as CO_2 from respiration. My proposal is to develop ships pushing skimmers up to a kilometre wide that harvest the plant material (just as whales harvest krill) which is then compressed into one metre cubes and also stored. These skimming ships also seed the ocean behind them with iron salts so that the plankton and other plants are quickly replaced by blooming due to the increased nutrition made available. I estimate that u million tonne of plankton material can be skimmed off the oceans and stored. This would translate to approximately v million tonnes of carbon per annum. As an alternative to storing this material, it might readily be 'fed' to a proportion of our symbiotic ant and bee farms to be readily converted to honey, wax and formic acid.*

4.   *Futureland*

*This is the title of Chapter Five of the prequel to this book titled "Nuclear Islam and Other Stories". In this chapter I outline the various scenarios of some hypothetical nation from the present until the year 2075. The sweetest of them was that nation that did **not** go down the 'nuclear fission' road but set in place sensible alternative technologies with the environment as the key point to a safe and healthy country for its citizens. A main target was the reclamation of land purely for parks and wilderness and the minimal proportion to be 25% of the total land area for the nation. This would create a new* *carbon sink by natural means. The land may be regulated and managed to some extent but must allow for the natural sequestering of carbon. Thus it cannot be 'farmed' for timber resources. Such farms are to be included in the proportion allocated to agriculture generally, which was set at 66% of the total land mass of the country. To achieve these targets the madness of runaway urbanisation must be halted and reversed by several means outlined in the chapter. The poorest of land is more suited to factories and urbanisation. Of course these targets may not be possible for every nation but should be aspired to. Some nations might do better to compensate for those that cannot reach the targets. This strategy would contribute to s million tonnes per annum of sequestered carbon compared to the current t million tonnes.*

5.   *Carbon Glass and Diamonds are a Girl's Best Friends*

*With the advent of cleaner coal technology in the coal production of electricity efforts are being made to capture all the carbon dioxide emissions. But what to do with this CO_2? Some government and privately owned electric companies are forcing the gas deep underground where it is hoped it will remain for millions of years. But one must find suitable deep rock strata and, as with nuclear waste, there is no guarantee that this CO_2 will remain there. Murphy's Law is inevitable. My suggestion is that this unwanted carbon dioxide be compressed and transported to an adjacent location which is self-sufficient with solar and wind combination electricity production. At this site, water is electrolysed by this electricity to make oxygen and hydrogen gases. The hydrogen gas is used to reduce the CO_2 to pure carbon and water (which may then be recycled). The carbon is then compressed to graphite or better still, diamond. I will refer to this synthetic material as **carbon glass** so that we know where it has come* from even though, in truth, it is diamond. One might say we are turning coal into diamonds! To protect the diamond and jewellery industry this 'carbon glass' is to be sequestered as part of the nation's carbon sequestering program and again stored in a safe place. Thus we can add to our equation: p tonne of coal producing q tonne of CO_2 to eventually produce r tonne of carbon glass. As we all know, diamonds are a girl's best friend but as a concentrated form of carbon this is the **best** form of carbon sequestration!

[I have made it most clear that I do not support nuclear fission as an energy source, however to safe-guard the carbon glass, one could combine it with nuclear waste and store in a stable and remote location]

CARBON DIOXIDE TO DIAMOND PLANT

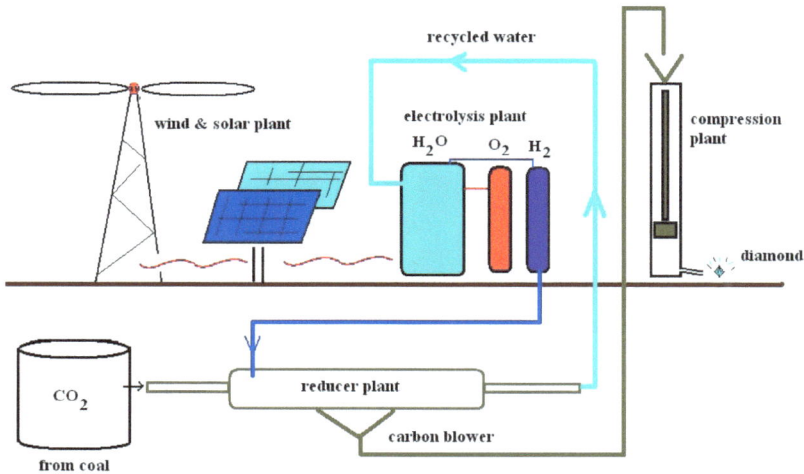

We have nice sunny desert areas in various parts of the planet. A hundred square kilometre area in Australia say could handsomely remove CO_2 from the atmosphere now by means of converting to carbon glass whilst we slowly phase out dirty coal power stations. The creation of solar power and wind power hardware should also be *manufactured at sites where the total energy of the plant/ factory complex is from the same technology. This is called bootstrapping and though expensive to begin with ensures a clean and viable future.*

Taking together my ideas presented in the five actions above we arrive at total carbon sequestered worldwide given by:

$$\Sigma = x + z + v + (s - t) + r$$

I am confident that the summation can easily meet the requirement of tonne per annum by Sir Michael Branson and his panel. In time, the value of $(s - t) + r$ that is increased forested areas plus carbon glass production, will suffice alone.

Selection of Most Suitable Compounds for Sequestering Carbon

Below I give the name, formula and percentage by weight of carbon contained therein to assist us in selecting those materials that might be sequestered or stored indefinitely to ensure the reduction of carbon in the atmosphere. The ant, the bee and miniscule creatures of the sea can make the difference. Abandoning nuclear fission as soon as possible is desirable and wind, solar, geothermal and gravipotential energy alternatives put in place as soon as possible. It cannot be achieved overnight and many Federal Governments around the world are reluctant to come into confrontation with their global masters of big corporations. But Big Government can be circumvented by local government and small communities. Each raindrop leads to a small cataract which leads to a river which leads to the sea. Soon alternative technologies will flood the planet to kick out the bad old habits and three cheers to all those brave governments that have seen the light and have made a start. From the table you can see that there are many compounds with very high percentages of carbon for

153

storage. The problem remains that many of these are valuable fuels which we need for our guzzling cars and other machines.

It's as though we all live on planet Chocolate, we all eat chocolate with the result of a chocolate atmosphere as a consequence which will eventually bring about the death of everyone! Decisions must be made and our habits and life actions must alter for salvation almost in an evangelical charismatic way. Sequestering these materials must become a discipline taken up by all nations working together. We must ensure the reduction of carbon in the atmosphere. The ant, the bee and the miniscule are the solution!

There is one final ingredient I have not yet mentioned and that is the water engine to replace all engines everywhere i.e cars, pumps, generators etc.

I know I am repetitive, but the legacy of nuclear waste spread around the planet will prove to have far greater impact and consequences on future generations than even global warming. Nuclear waste is a formula ringing the death knell of a safe and healthy environment which is a contradictory view to its proponents. Mathematically, we cannot escape the eventual effects of heavy radioactive isotopes working their way through the global environment. Its proponents hide their minds from this inevitable truth and continue to propagate a proliferation of lies to save their multi trillion dollar industry at the expense of future generations of human lives. They are heartless and just don't care. Of these four mining industries in Australia: coal, asbestos, lead and uranium; uranium mining is the most detrimental to the health of workers and to their families living in close proximity to the mining area. Lead dust is also accumulative in the human body

throughout one's life and causes serious health problems such as
dementia (Mad Hatter Disease)

Percentage Carbon per Carbon Con taining Compound

Name of Compound	Molecular Formula	Empirical Formula	% carbon by weight	comment	state
carbon	C	C	100		solid
carbon monoxide	CO	CO	43	contains oxygen	gas
carbon dioxide	CO2	CO2	27	contains oxygen	gas
carbonic acid	HCO3	HCO3	19	contains oxygen	liquid
hydrogen cyanide	HCN	HCN	44	highly toxic	gas
acetyline	CH3CN	C2H3N	58.5		gas
benzene	C6H6	CH	92		liquid
cyclohexane	C6H12	CH2	86		liquid
methane	CH4	CH4	75		gas
ethane	C2H6	CH3	80		gas
propane	C3H8	C3H8	82		gas
bees wax	C24H50	C12H25	85		solid
methanol	CH3OH	CH4O	37.5	contains oxygen	liquid
ethanol	C2H5OH	C2H6O	52	contains oxygen	liquid
formic acid	HCOOH	CH2O2	26	contains oxygen	liquid
acetic acid	CH3COOH	C2H4O2	40	contains oxygen	liquid
bees bodies	0.5(HCOOH.CH3COOH)	C3H6O4	34	contains oxygen	liquid
propanoic acid	C2H5COOH	C3H6O2	48.5	contains oxygen	liquid
butanoic acid	C3H7COOH	C4H8O2	54.5	contains oxygen	liquid
sugar	C6H12O6	CH2O	40	contains oxygen	solid
starch	C6H10O5	C6H10O5	44.5	contains oxygen	solid
formaldehyde	HCHO	CH2O	40	contains oxygen	liquid
acetaldehyde	CH3CHO	C2H4O	54.5	contains oxygen	liquid
propanal	C2H5CHO	C3H6O	62	contains oxygen	liquid
butanal	C3H7CHO	C4H8O	66.5	contains oxygen	liquid
dicyanide	NCCN	C2N2	46	highly toxic	gas
ethene	C2H4	CH2	86		gas
propene	CH3CHCH2	CH2	86		gas
propanediene	CH2CCH2	C3H4	90		gas
butene	CH3CH2CHCH2	CH2	86		liquid
butanediene	CH2CHCHCH2	C2H3	89		liquid
carbon disulphide	CS2	CS2	16		liquid
silicon carbide	SiC	SiC	33		solid
diamond	(C)n	C	100	exotic	solid
graphite	(C)n	C	100		solid
coal	C+S+N+P+H2O	C+ impurity	93	average	solid
cyclopropanetriene	C3	C	100	exotic	liquid
cyclobutanetetrene	C4	C	100	exotic	liquid

6. The Water Engine.

The schema below shows in principal how the water engine works. Of course there needs much refinement and the key is in the design of the engine that either produces mechanical energy in a conventional fashion or produces electrical energy directly. Much research is going on in this area. What the world needs is a simple 'take anywhere' design that will provide immediate mechanical or electrical energy. Like a battery, the oxygen and hydrogen need to be regenerated. This is most conveniently done by a solar panel whilst not in use or plugged in to the mains if immediate recharging is necessary. The solar panel does not have to be part of the device but can be the combined total surface of a car or the roof of the garage etc. How the

PRINCIPAL OF WATER ENGINE

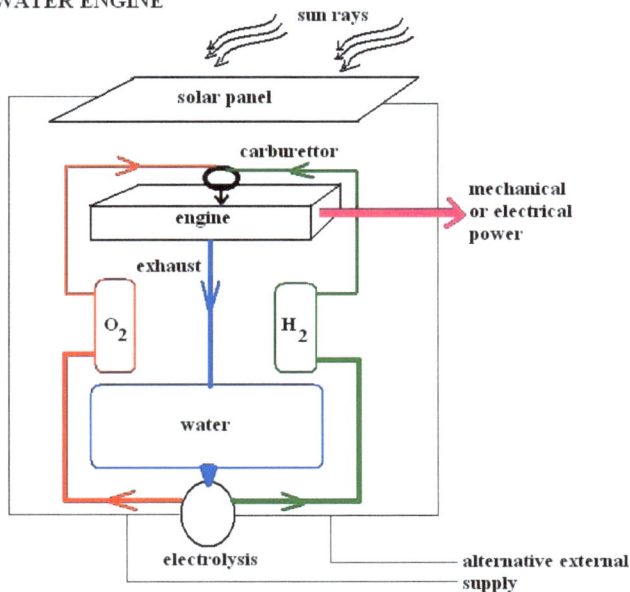

gases are mixed in the engine needs much thought with the design of valves to ensure the complete safety. I envision a totally sealed system with all components contained inside a block that is easy to transport and easily exchanged as a total sealed unit. It may be applied to do mechanical work directly or act as a generator to provide limited electrical power. The engine might be mechanical or, more efficiently, be operated as a fuel cell.

Let us not allow lassitude to weigh down upon us but rather develop alternatives now. We cannot sit and watch as our hydrocarbon fuels dwindle bringing us the fate of the lemming. Anything that reduces carbon emissions now will assist in a cleaner atmosphere and stave off the drama of a global nightmare!

7.   Population

Early in this chapter I made a supercilious remark regarding reducing the world population as a solution. However we need to face mathematics and not enough governments have serious long term plans to overcome their burgeoning populations. It is said that the world's population doubles every thirty years. Let us take a parsimonious view here and say 33 years for the next two doublings. If the population in January 2010 was 6.45 billion souls , this means 12.7 billion by 2042 and 25.4 billion by 2075. Thus in just 66 years we can expect a four-fold increase in the world's human population. If we consider a rocketing of technology and modernisation in third world countries, particularly in Asia, is it not obvious that the problem of carbon emissions will remain even with all the best intentions and alternative measures set in place? As stated in the prequel (Nuclear Islam and Other Stories) crunch point occurs around this date (2075) for the availability of resources (land, clean air, clean water, energy and food).

Many of my contemporaries have scoffed at my dire predictions and some are still chuckling into their sleeve. But many others have given my predictions serious thought and many red faces have appeared with fear on their brow and a trembling of the lips. We are not looking at centuries ahead... just a

157

handful of decades! A country such as Indonesia with a growing scarcity of land and rice paddies to feed its 230 millions in 2009 will have a population of at least 920 billion by 2075. It will probably burst through the one billion mark due to its predominant poor lower class and the Islamic culture of many children (it is not uncommon to see families with six to ten children). This country must attend more to its population crisis to avert a humanitarian catastrophe. It must set in place strict quotas on family size now.

So generally, nations must look to methods of curtailing population growth and even achieve lower populations if at all possible. With the rate of expansion and development of modern technology coupled with its culture of both necessity and waste it would be favourable for the world to return to those population levels of 1952 (around 1.65 billion for the whole world). But this is unlikely to happen.

Estimates of the Sources of Greenhouse Gases by %

respiration land plants	4	**TABLE OF PERCENTAGES OF GREENHOUSE GAS EMISSIONS BY SOURCE**
respiration land animals	2.5	
respiration ocean plants and animals	8	
manufacture iron and steel	4.5	The table includes all those gases such as CO_2, methane, oxides of nitrogen, CFC's etc. generally considered to contribute towards a heating of the troposphere (enhanced greenhouse effect)
manufacture concrete	3.5	
exhaust cars, trucks & motorcycles	9.5	
industrial generators	1	
coal fired power plants	20	
nuclear power industry	2	
mining industries	3	note: this table excludes water as a source even though it is also a greenhouse gas.
armaments manufacture(all)	2	
petrochemical refineries	3.5	
all other manufacturing and industry	9.5	
domestic & international aircraft	1.5	
war zones	1.5	
farming practices	4	
decay land, lakes, swamps	4.5	
decay oceans	7	
annal exhalation humans	2	
annal exhalation animals	3.5	
forest fires	2	
volcanic and plutonic action	1	

This chapter is more about the consensus of the scientific world on climate change and global warming. As you will find, there remains no strong commitment of scientists (other than those shamefully seeking funding) on whether humans are responsible for any sudden or significant changes with regards to global climate. Climate is the average of weather conditions over long time periods. It is a complex system that is constantly changing and it is important to note that individual weather events cannot be taken to be indicative of trends; neither locally nor on a global basis. Computer modelling, while improving with the sophistication of software and the advance in computing power of machines, is not an exact science for the predicting of weather changes. Scientists examine data from a variety of sources to build a picture of climate change over thousands of years. Ice cores from glaciers and polar ice, growth of tree rings, pollen in sediments and other pointers have contributed to this modelling. However, graphs of data from these sources build on averages over time intervals of centuries. It is impossible to interpolate to provide accurate temperatures for individual years or even means for time periods as short as a decade. Using these techniques, climate scientists have a picture of mean global temperatures for the past 650000 years and this is being extended from time to time as new improved data comes to hand. A difficulty with the present warming is that temperatures between 1100 and 1300 AD were very similar to those being experience at present. Further, the data of the past indicates a lag between temperature rise and the rise in CO_2 atmospheric level. This poses the awkward question: "is *temperature rise a function of increasing CO_2 level or is increasing CO_2 level a function of temperature rise?" Unfortunately for the environmentalist's stance, the data from the past suggests that the latter of these two seems to be the truth.*

The other problem in sorting out what is happening just at present is that, due to the enormous size and complexity of the global system, it generally takes up to one hundred years before we see any global change from factors or occurrences causing these changes. An instance is that if the mean temperature of the oceans were to rise by 0.5 of a degree Celsius say, the effect on glaciers, Greenland and the West Antarctic ice shelf could take hundreds of years. A total melt would certainly take in excess of 1000 years. As we are a considerable time from the end of the last ice age, the residual ice on the Earth, if it were to melt completely, would result in a sea level rise of no more than 5.75 metre. A further small rise of up to 0.25 metre might be attributable to thermal expansion of the oceans. Thus the maximum rise possible is around 6 metre and unlikely over the next 1000 to 5000 years. Current estimates predict a median rise of just under 0.5 metre over the next 100 years.(Having said this, even a 0.5 metre rise would have significant impacts on Pacific islands, Bangladesh and other low lying parts of nations around the world.)

The Earth has a natural Greenhouse Effect due to the various molecules in the atmosphere that permit the transmission of shorter wavelengths (including visible light) to reach the surface of the planet but absorb a small proportion of the longer infrared wavelengths from energy radiated back from the surface towards space. Among these molecules that absorb this energy is carbon dioxide which makes up about 0.03 % of the atmosphere. The prime absorber is water molecules. Of course oxides of nitrogen and sulphur also contribute to some absorption. Primarily then, the Earth and its atmosphere radiate long wave infrared due to its surface temperature which is attenuated by the presence of greenhouse gases.

Scientists have been measuring atmospheric CO_2 directly since the 1800s. The pre-industrial concentration of CO_2 was 280 parts per million whereas the 2005 level was recorded as 380 parts per million. There have been various claims of localise phenomena which have later been extrapolated to the whole world. This cannot be done as it has no scientific basis. Of course we have seen the melting of glaciers at various locations but the evidence points to a trend that has been present for up to two hundred years. Increase in population and denuding of local forests can explain some local decline in glaciers. In other parts of the world, notably Tibet, there has been an increase in ice volume associated with the mountain peaks and glaciers. Glacier retreat in Glacier National Park, USA, has been an ongoing phenomenon since the 1850s and so it is difficult to place the blame on human activity. All glaciers are losing mass to melting and gaining mass from precipitation. Whilst it appears that losses are currently outpacing gains for most, as we can see this is not the case for all. It is a fact that Himalayan snow accumulation has steadily declined since the mid 19th Century. When we consider that the Medieval Warm Period temperatures are very similar to those of today, again we cannot state emphatically that the current picture has an anthropogenic cause.

The UK Climate Research Unit gives 1998, 2002 and 2005 as the hottest years in the last 100 years. However the data describes a post 1970 warming bias which may have various causes, the urban island

161

heat effect being just one. Climatologists note that it is too early to distinguish between a trend and an anomaly. We must never apply extrapolation of a local effect to the global situation. For the USA, data from 48 states lists the hottest years (from hottest to coolest) as: 1934, 1998, 1921, 1931, 1999, tied 1953 1990 and 2001, tied 1987 and 2005 with the mean difference of 0.4 ^0C between 2005 and 1934.

Global Circulation Models (GCMs) on the best of our computers still fail to replicate observed temperature changes based on past data and all the factors currently understood to influence climate. Factors not yet understood are incorporated with particular values, but as said, GCMs fail to make predictions that can be subsequently verified. Natural variability is inherent in these models making human interference difficult to pinpoint with any degree of certainty.

On the question of recognised recent peaks in hurricanes, again we cannot be certain if these are part of some new trend due to global warming or a result of some other natural cycle. Increases in North Atlantic tropical storm activity has been offset in observed decreases in Northeast Pacific tropical storm activity leading to minimal change globally. The picture is similar for tornadoes; in fact the increase in the sensitivity of scientific instrumentation has permitted the recording of lesser tornadoes that would otherwise have been unrecorded several decades ago. A single occurrence such as the horrendous hurricane Katrina, whilst important in our life-time, cannot be linked to any climate trend.

How often have we seen 'floods' on the news with some commentator describing then as 'the worst in living memory blah blah blah' The facts are that many of these are due to increased population particularly in flood prone areas along with land use changes. There seem to be heavy rain events in some locations and less in others but overall no increase in the frequency of these. Studies on flooding of the Elbe and Oder rivers in Europe dating back to the twelfth Century realised no new trend on flood occurrence. A similar study for the Yangtze Delta in China determined the greatest frequency of extreme flooding to have occurred between 1500 and 1700 AD, during the transition period between the Medieval Warm Period and the Little Ice Age. Extremes such as drought and flood form part of a natural aspect of what is termed a 'chaotic climate

system'. No modern trends indicate that there is an increase in either of these phenomena on a global basis.

What seems to be certain is that levels of CO_2 gas are increasing in the atmosphere and that this is likely to continue with the rapid development of the two super economies of China and India. As *described earlier, even with the percentage of coal as part of total energy production decreasing, both these countries will be seen to increase their coal consumption dramatically over the following decades (for clarification look at Table 4.1 and Graph 4.1).*

Whether we can label the decrease in the Arctic ice by changed mean global sea temperature is also impossible. Some scientists argue that it is due to changed conditions of ocean currents and wind. It is undeniable, however, that the extent of the Arctic ice is much reduced and continuing to decrease in total area. It has been predicted that the Northwest Passage will soon be navigable by commercial shipping, knocking off some 7000 Km between the west coast of America and Europe. "Ice has retreated to about three million square kilometres", Leif Pedersen of the Danish National Space Centre, said in the statement (September 2007). From satellite pictures, the European Space Agency said the previous low was four million square kilometres back in 2005. There is doubt over when such a route will become safe and practical.

Between 34 and 15 million years ago (Mya), when planetary temperatures were 3–4 °C warmer than at present and atmospheric CO_2 concentrations were twice as high as today, the Antarctic ice sheets may have been unstable. Oxygen isotope records from deep-sea sediment cores suggest that during this time fluctuations in global temperatures and high-latitude continental ice volumes were influenced by orbital cycles but it has hitherto not been possible to calibrate the inferred changes in ice volume with direct evidence for oscillations of the Antarctic ice sheets. Sediment data from shallow marine cores in the western Ross Sea exhibit well dated cyclic *variations, and link the extent of the East Antarctic ice sheet directly to orbital cycles during the Oligocene/Miocene transition (24.1– 23.7 Mya).*

www.nature.com/nature/journal/v413/n6857/abs/413719a0.html

There have been no significant net changes in the ice volume of the East Antarctic in recent years with research suggesting it is growing. Whereas there has been some outflow from the West Antarctic ice sheet and from Greenland, the consensus remains at thousands of years for any significant outflow. During the past ten million years (Mya) we see that during the early Pliocene West Antarctic ice- volume was reduced by 70%. During the late Miocene the ice-sheet was highly dynamic with frequent advances and retreats. The upper Messinian to early late Pliocene was a time of repeated ice-sheet collapses. Starting at 3.2 Mya the ice-sheet became a permanent feature, occupying the shelf during most of the glacial half cycles.

Interesting to note here that a 2003 survey by Bray of 530 climatologists were asked "To what extent do you agree or disagree that climate change is mostly the result of anthropogenic causes?" on a scale of 1 to 7, 56% agreed slightly to strongly with 9% of respondents saying they strongly agreed. The superficial study was generally rejected on the grounds of obligatory acknowledgement of "climate change" for researchers wishing to get their papers published. Many scientists hold sincere disagreements with presentations of evidence of the global warming hypothesis based on examination of evidence. We have seen smear campaigns by both the energy lobbyists as well as the environmentalists neither of which has assisted the scientific study of climate and climate change on a global basis. Whilst Mr Gore's book has brought to the attention of the public at large the possible consequences of global warming, much of what he has to say is not backed by transparent data. Many of his interpretations of single events fall into the category of taking the extreme point of view and grandiose sensationalism. Having said this, the author feels that Mr Gore's book is valuable in that it brings a greater awareness of the extravagance of the developed nations regarding energy consumption and the potential for real environmental harm to the planet if we continue in such vein. The author's viewpoint is that the sheer weight of human population is the greatest threat to our future with little attention by governments the world over to address the problem. The Earth itself has its own solution but what can we do to solve this problem?

Lastly, in the past we have seen dramatic changes in the climate record due to sudden increases in dust, oxides of sulphur and oxides of nitrogen. These dramatic events have been around the same time as massive volcanic eruptions and lava flows at various sites upon the Earth. A nuclear exchange between major nations would most likely simulate these earlier natural disasters followed by global cooling in what has been described as a "nuclear winter". This scenario would certainly then fit the bill of "human activity causing dramatic climate change".

$$6H_2O_{(l)} \; + \; 6CO_{2(g)} \; \rightarrow \; C_6H_{12}O_{6(aq)} \; + \; 6O_{2(g)} \quad \Delta H = + 2800 \; kJ.mol^{-1}$$

water   carbon dioxide   glucose   oxygen   energy absorbed

This is the famous equation studied by all high school students in junior science. It is the equation for photosynthesis: from the reactants water and carbon dioxide and in the presence of sunlight, green leafed plants are able to produce glucose and oxygen.

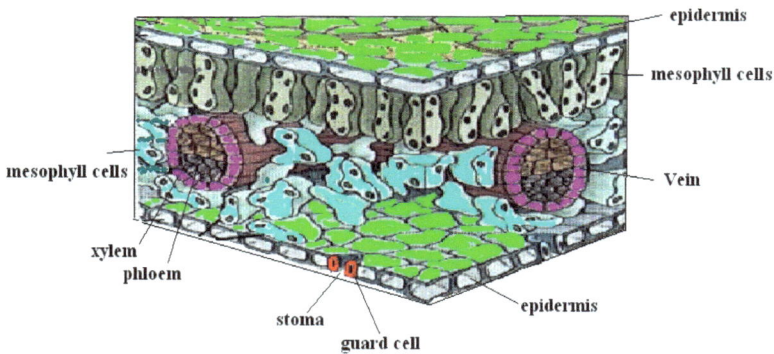

Here we see a cross section of a leaf, showing the anatomical features important to the study of photosynthesis: stoma, guard cell, mesophyll cells, and vein. [adapted from Purves et al., "Life: The Science of Biology", 4th Edition, by Sinauer Associates and WH Freeman.]

Water is absorbed through the membrane of the roots in a process called osmosis. Then it is transported up to the leaves through specialized plant cells called xylem . Plants must prevent drying out to achieve this have evolved specialized structures known as stomata to allow gases to enter and leave the leaf. Carbon dioxide cannot pass through the cuticle, the protective waxy layer covering the leaf, but it can enter the leaf via the stomata, small holes flanked by two guard cells. Oxygen produced during photosynthesis can only pass out of the leaf through these stomata. Now while these gases are moving between the inside and the outside of the leaf, a great deal water is also lost, with some species losing up to 400 litres of water per hour during extremes of heat. Carbon dioxide may enter single-celled and aquatic autotrophs through no particular specialized structures.

In the chloroplasts of the leaf cells we have the substance chlorophyll which is the catalyst for the photosynthesis chemical reaction. Chlorophyll, like a green pigment, is common to all photosynthetic cells, and absorbs all wavelengths of visible light except for green, a little blue and orange which it reflects to be seen by our eyes.

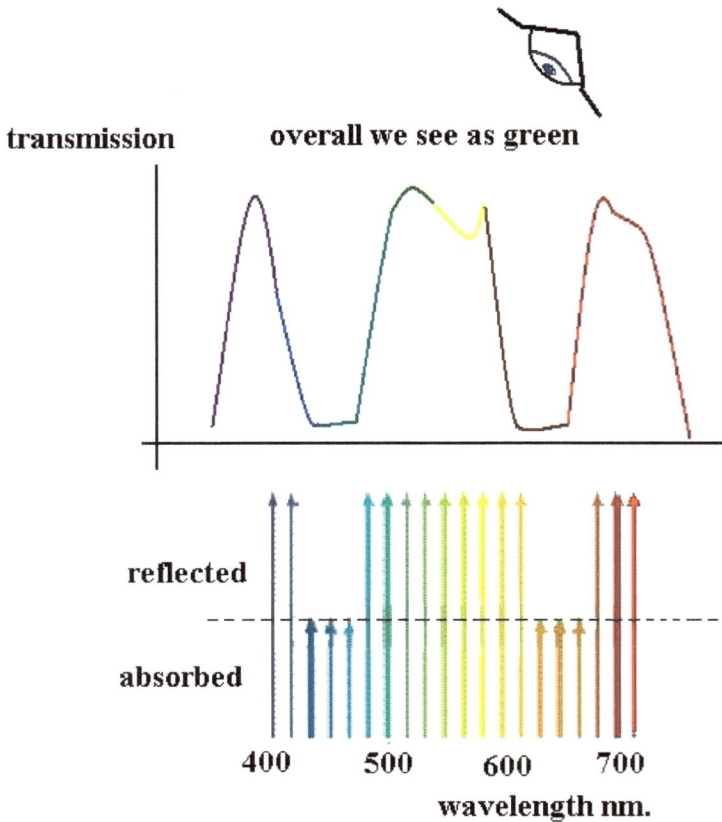

As we see from the following figure, visible light is only a small part of the total spectrum of electromagnetic waves.

Not only is Chlorophyll a complex molecule, but we see several slight variations of structure of chlorophyll occurring among plants and other photosynthetic organisms. All photosynthetic organisms i.e plants, certain protistans, prochlorobacteria, and cyanobacteria have chlorophyll A. Additional pigments absorb energy that chlorophyll A does not absorb. These include chlorophyll B, with C, D, and E found

The Electromagnetic Spectrum

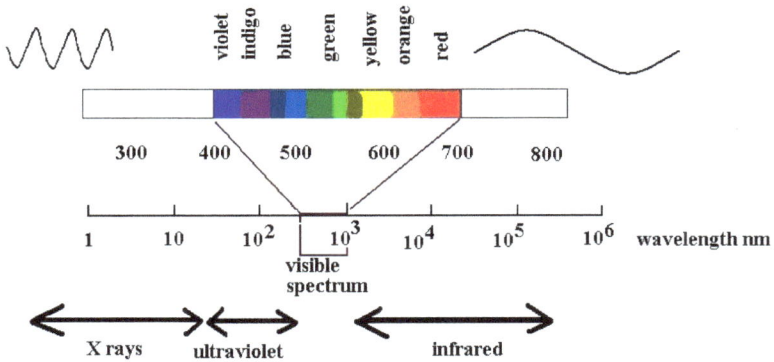

in algae and protistans. Also we have xanthophylls, and <u>*carotenoids*</u> *such as beta-carotene. Chlorophyll A absorbs its energy from the Violet-Blue and Reddish orange-Red wavelengths, and a little from the intermediate Green-Yellow-Orange wavelengths.*

Carotenoids and chlorophyll B absorb some of the energy in the green wavelength. But why not so much in the orange and yellow wavelengths? Both chlorophylls also absorb in the orange-red end of

170

the spectrum having longer wavelength and lower energy. The origin of photosynthetic organisms in the early oceans may account for this. The shorter (and more energetic) wavelengths do not penetrate much below 5 meters from the surface of sea water. The ability to absorb some energy from the longer and more penetrating wavelengths might have been an advantage to early photosynthetic algae that were not able to reside continuously in the upper photo zone of the ocean.

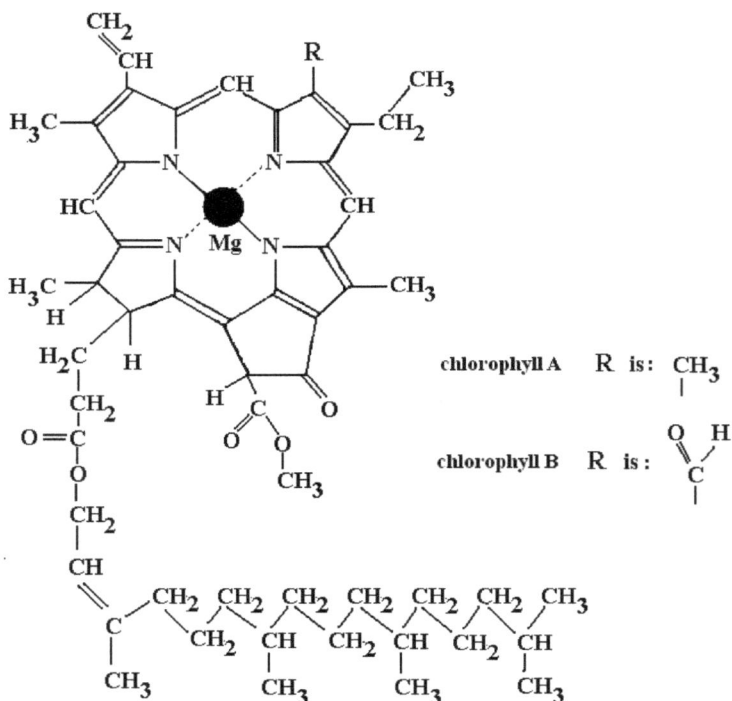

chlorophyll A   R is: CH_3

chlorophyll B   R is: O H \ / C |

The Chlorophyll Molecule

It is found that energy may trigger many types of chemical reaction, photosynthesis being an example. Chlorophyll only triggers a chemical reaction when it is associated with proteins embedded in a membrane such as in a chloroplast or the membrane infoldings found in photosynthetic prokaryotes such as cyanobacteria and prochlorobacteria.

Absorption Spectrum for Some Plant Pigments

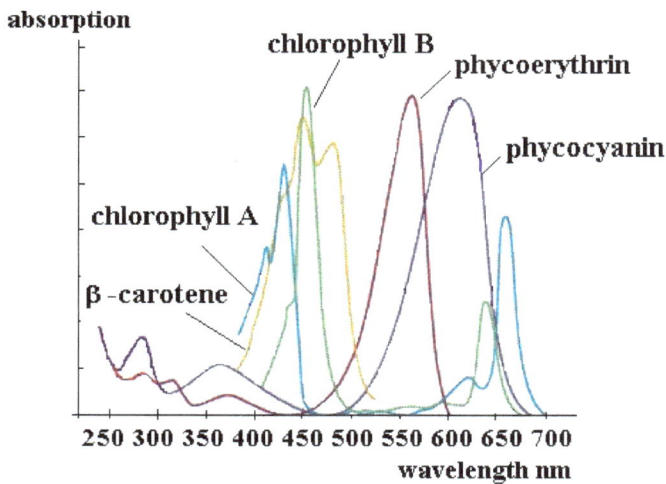

The structure of the chloroplast and photosynthetic membranes

The thylakoid is the structural unit where photosynthesis takes place. Both photosynthetic prokaryotes and eukaryotes have these flattened vesicles containing photosynthetic chemicals. However, only eukaryotes have chloroplasts with a surrounding membrane. Thylakoids are stacked like pancakes in stacks known collectively as grana. The areas between grana are referred to as stroma. While the mitochondrion has two membrane systems, the chloroplast has three, forming three compartments.

The stages of photosynthesis are twofold. The first process is the 'light dependent process' which requires the direct energy of light to make energy carrier molecules that are used in the second process. The 'light independent process', also termed the

Structure of a Chloroplast

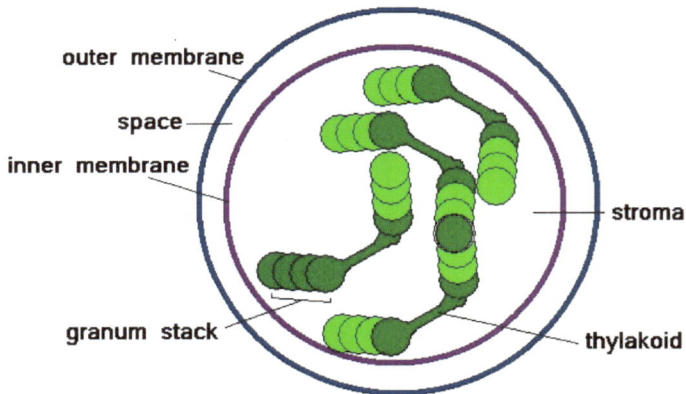

'dark reaction', occurs when the products of the light reaction are used to form carbon to carbon C-C covalent bonds of carbohydrates

In the light dependent process, light strikes chlorophyll A in such a way as to excite electrons to a higher energy state. In a series of reactions the energy is converted, via an electron transport process, into ATP and NADPH. Water molecules are split in the process, releasing oxygen as a by-product of the reaction. The ATP and

NADPH are used to make the C-C bonds in the light independent process (or dark reaction).

In the light independent process, carbon dioxide from the atmosphere (or water for aquatic/marine organisms) is captured and modified by the addition of hydrogen to form carbohydrates whose general formula is: $[CH_2O]_n$. The incorporation of carbon dioxide into organic compounds is known as carbon fixation. The energy for this comes from the first phase of the photosynthetic process. Living systems cannot directly utilize light energy, but can, through a complicated series of reactions, convert it into C-C bond energy that can be released by glycolysis and other metabolic processes.

The Dark Reaction

Carbon-fixing reactions are also known as the dark reactions (or light independent reactions). Carbon dioxide enters single-celled and

aquatic autotrophs through no specialized structures, just simply diffusing into the cells. Land plants must guard against drying out and so have evolved specialized structures known as stomata to allow gas to enter and leave their leaves. The 'Calvin Cycle' occurs in the stroma of chloroplasts. Carbon dioxide is captured by the chemical *ribulose biphosphate or RBP. RBP is a five carbon chemical molecule. Six molecules of carbon dioxide enter the Calvin Cycle, eventually producing one molecule of glucose. The reactions in this process were worked out by Melvin Calvin*.*

*[One of the new areas, cultivated both at Donner and the Old Radiation Laboratory, was the study of organic compounds labelled with the isotope carbon-14. Melvin Calvin took charge of this work in the late 1940's in order to provide raw materials for John Lawrence's researches and for his own study of photosynthesis. Using carbon-14, available in plenty from the Hanford reactors, and employing the new techniques of ion exchange, paper chromatography, and radio-autography, Calvin and his many associates mapped the complete path of carbon in photosynthesis. This accomplishment earned him the Nobel Prize for Chemistry in 1961.]

I give some detail of the first step in the Calvin Cycle below. For a more detailed account, readers should look to a reference such as: Purves et al.," Life: The Science of Biology", 4th Edition, by Sinauer Associates and WH Freeman.

(www.whfreeman.com and www.sinauer.com)

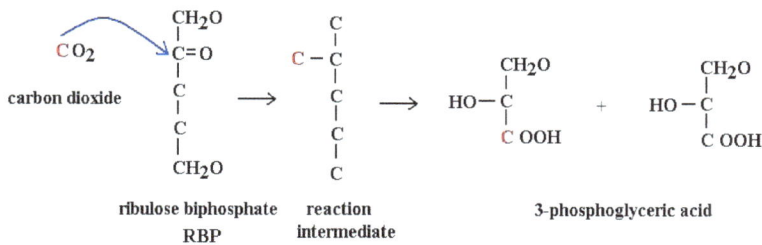

CO₂ carbon dioxide → ribulose biphosphate RBP → reaction intermediate → 3-phosphoglyceric acid

The First Steps in the Calvin Cycle.

The first stable product of the Calvin Cycle is phosphoglycerate (PGA), a three carbon chemical molecule. The energy from ATP and NADPH energy carriers generated by the photosystems is used to attach phosphates to (phosphorylate) the PGA. Eventually there are 12 molecules of glyceraldehyde phosphate, also known as phosphoglyceraldehyde or PGAL, two of which are removed from the cycle to make glucose. The remaining PGAL molecules are converted by ATP energy to reform 6 RBP molecules, and thus start the cycle again. Every reaction in this process is catalyzed by a different reaction-specific enzyme.

Some plants have developed a preliminary step to the Calvin Cycle While most carbon fixation begins with RBP, the C-4 cycle begins with a new molecule, phosphoenolpyruvate (PEP), a three carbon chemical molecule that is converted into oxaloacetic acid, OAA, when carbon dioxide is combined with PEP. The OAA is converted to Malic Acid and then transported from the mesophyll cell into the bundle-sheath cell, where OAA is broken down into PEP plus carbon dioxide.

176

The carbon dioxide then enters the Calvin Cycle, with PEP returning to the mesophyll cell. The resulting sugars are now adjacent to the leaf veins and can readily be transported throughout the plant. Thus C-4 photosynthesis involves the separation of carbon fixation and carbohydrate synthesis.

There exist distinct anatomical differences between C3 photosynthesis and C4 photosynthesis leaves.

Respiration

The overall reaction for cellular respiration is:

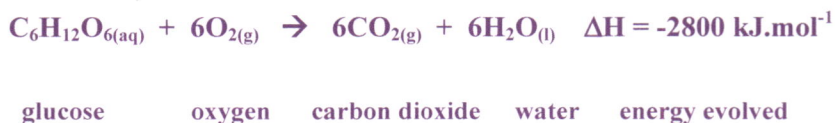

$$C_6H_{12}O_{6(aq)} + 6O_{2(g)} \rightarrow 6CO_{2(g)} + 6H_2O_{(l)} \quad \Delta H = -2800 \text{ kJ.mol}^{-1}$$

    glucose   oxygen   carbon dioxide   water   energy evolved

The complexity of respiration can be studied further in many of the more academic texts in cellular biology and will not be described in detail here. From the first and simplest one celled living things to the complexity of trees, reptiles and mammals, when cells burn carbohydrates such as glucose to produce energy, carbon dioxide is a product. Plants lose CO_2 via their stomata; humans and mammals exhale the same via the lungs.

Back to plants: In glycolysis and the Krebs cycle*, there are electrons released as the glucose molecule is broken down. The cell must deal with these electrons in some way, so they are stored by forming a compound called NADH by the chemical reaction:

$$NAD^+ + H^+ + 2e^- \rightarrow NADH.$$

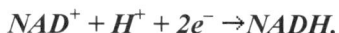

This NADH is used to carry the electrons to the 'electron transport chain', where more energy is harvested from them.

In eukaryotes, the pyruvic acid from glycolysis must be transferred into the mitochondria to be sent through the Krebs cycle, also known as the 'citric acid cycle', at a cost of one ATP per molecule of pyruvic acid. In this cycle the pyruvic acid molecules are converted to CO_2, and two more ATP molecules are produced per molecule of glucose.

*[The Krebs Cycle: "in all plants and animals a series of enzymatic reactions in mitochondria involving oxidative metabolism of acetyl compounds to produce high-energy phosphate compounds that are the source of cellular energy" discovered by Hans Krebs]

Many of the compounds that make up the electron transport chain belong to a special group of chemicals called cytochromes. The central structure of a cytochrome is a porphyrin ring like chlorophyll but with iron in the centre (as we saw above, chlorophyll has magnesium). A porphyrin with iron in the center is called a heme group, and these are also found in hemoglobin in human blood.

Finally in the electron transport chain, the used up electrons, along with some hydrogen ions are combine with oxygen to form water as a waste product:

$$4e^- + 4H^+ + O_2 \rightarrow 2H_2O$$

178

The spanners and other tools in cellular bio-chemistry are complex molecules called enzymes, comprising of arrays of proteins.

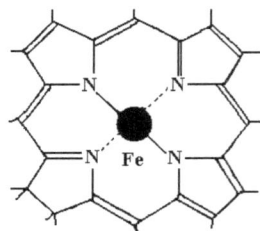

The Porphyrin Ring of the Heme Group

Many of these enzymes in the cells of organisms need additional helpers to function. These non-protein enzyme helpers are called cofactors and can include elements such as iron, zinc, or copper. If a cofactor is an organic molecule, it then is called a coenzyme. Many of the vitamins we require are used as coenzymes to help our enzymes to do their jobs. For example, Vitamin B_1 (thiamine) is a coenzyme applied to remove CO_2 from various organic compounds.

Without enough of these B vitamins, we would not be able to readily obtain energy from food. B_6 (pyridoxine), B_{12} (cobalamin), pantothenic acid, folic acid, and biotin are all other B vitamins which serve as coenzymes at various points in metabolizing our food. B_{12} has cobalt in it, a mineral which we need in only very minute quantities (thus described as a trace element), but whose absence can cause anatomical symptoms of deficiency.

These notes on respiration I admit are sketchy and as mentioned earlier the serious reader should look to the reference material for a deeper insight. But then you could do that for so many of the topics glossed over in this writing.

179

Eleven   **Planet Ocean**

Image (AS17-148-22727) courtesy of the Image Science & Analysis Laboratory, NASA Johnson Space Center

It has been said that our planet would be more appropriately named planet Ocean than Earth and looking at the Apollo 17 pic, its not hard to understand why!

More than three fifths (71% actually) of the surface of the Earth are covered by oceans namely, the Arctic, the Atlantic, the Indian, the

Pacific and the Southern totalling some 335000 square kilometres. Looking at the ocean facts data it's interesting to note the following:

The Antarctic Ice Sheet is almost twice the size of the United States

Nearly all volcanic activity (90%) occurs under the ocean

The highest tides in the world are at the Bay of Fundy, New Brunswick/Nova Scotia and the difference between high and low tide being as great as 18 metres

The longest mountain range is the Mid-Ocean Ridge, which winds around the globe from the Arctic Ocean to the Atlantic, skirting Africa, Asia and Australia, and crossing the Pacific to the west coast of North America. It is four times longer than the Andes, Rockies, and Himalayas combined

El Niño, a periodic shift of warm waters from the western to eastern Pacific Ocean, has dramatic effects on climate worldwide

At the deepest point in the ocean the pressure is about1.25 tonnes per square centimetre

The temperature of almost the entire deep ocean is only 4^0 Celsius

If it were easily gotten, all the gold in the world's seawater would provide each person on Earth with more than 3 kilogram.

The world's tallest known iceberg was measured off western Greenland to be 175 metre

The inactive volcano of Hawaii, Mauna Kea stands above sea level, yet it is almost 10800 metres tall if measured from the ocean floor to its summit.

The largest recorded tsunami measured 65 metres above sea level when it reached Siberia's Kamchatka Peninsula in 1737.

A study by scientists at Bristol University reveals that the ocean has become more acidic than ever before, that is comparing acidy levels over the last 65 million years. The last time the acidity in the oceans rose there was a massive die off of the organisms that lived at the lowest depths.

Ocean acidity is a main concern of scientists studying climate change. Around half of all carbon dioxide produced by humans since the industrial revolution has dissolved into the world's oceans with adverse effects for marine life. This first comprehensive look at ocean storage of carbon dioxide found that the world's oceans serve as a massive sink that traps the greenhouse gas. The researchers say the oceans' removal of the carbon dioxide from Earth's atmosphere has slowed global warming.

But in a second study scientists say the sink effect is now changing ocean chemistry. The resulting change has slowed growth of plankton, corals, and other invertebrates that serve as the most basic level of the ocean food chain causing a severe impact on marine life generally.

Although performing a great service to humankind by removing this carbon dioxide from the atmosphere, there are potential consequences for the biology and ecosystem structure of the oceans. Furthermore, it is likely that there is a limit to this natural sequestering of carbon dioxide. As acidity levels increase, natural calcium carbonate in shells and deposits on the ocean floor will start to release carbon dioxide back to the atmosphere. Once commenced, this process will be virtually unstoppable. Today's level of atmospheric carbon dioxide is thought to be only half of what scientists have predicted atmospheric

182

levels should be, based on estimates that humans have contributed 244 billion metric tons of carbon dioxide to Earth's atmosphere over the last two hundred years.

The other half of the emitted carbon dioxide is thought to have been taken up by both the oceans and the land, but with the lion's share absorbed into the oceans. The data was collected by two international research programs: the World Ocean Circulation Experiment (WOCE) and the Joint Global Ocean Flux Study (JGOFS). The results suggest that the oceans have taken up 48 percent of all carbon dioxide emitted from fossil fuel burning, cement and other manufacture since 1800. It has been postulated that the amount of carbon dioxide the oceans have currently taken up is about a third of what they can hold. Once saturation has been reached, the researchers warn, the rate of global warming is likely to accelerate.

Because carbon dioxide is an acid gas, the surface ocean pH is dropping. The surface of the oceans, where most marine life is found, might soon become more acidic than it has been in over five million years.

The increase in acidity makes it difficult for shell-forming animals and some algae to amass carbonate ions from seawater to form their calcium carbonate shells.

Corals, some types of mollusc, and tiny planktonic organisms called foraminifers and coccolithophorids could all be affected. It should be noted that many of these species are key links in the marine food chain.

183

Past studies have shown that at atmospheric carbon dioxide concentrations of 700 to 800 ppm, (which some say could be reached by the end of this century) the rate at which these organisms can form shells could be reduced by as much as 45 percent. This could result in dramatic effects and alter food web structures in ways that cannot be predicted. Coral experts say that it is possible that prior to the industrial age, the world's oceans might have absorbed so much atmospheric carbon dioxide that the process played a role in limiting terrestrial plant growth.

The oceans contain about 50 times more CO_2 than the atmosphere and 20 times more than the land. CO_2 moves between the atmosphere and the ocean by molecular diffusion when there is a difference between CO_2 gas pressure (pCO_2) between the atmosphere and oceans. For example, when the atmospheric pCO_2 is higher than the surface ocean, CO_2 diffuses across the air-sea boundary into the sea water. The oceans are able to hold much more carbon than the atmosphere because most of the CO_2 that diffuses into the oceans reacts with the water to form carbonic acid and its dissociation products, bicarbonate and carbonate ions. The conversion of CO_2 gas into nongaseous forms such as carbonic acid and bicarbonate and carbonate ions effectively reduces the CO_2 gas pressure in the water, thereby allowing more diffusion from the atmosphere.

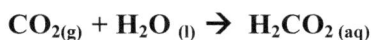

$$CO_{2(g)} + H_2O_{(l)} \rightarrow H_2CO_{2\,(aq)}$$

$$H_2CO_{2\,(aq)} \rightarrow H^+_{\,(aq)} + HCO_3^-{}_{\,(aq)}$$

$$HCO_3^-{}_{\,(aq)} \rightarrow H^+_{\,(aq)} + CO_3^{2-}{}_{\,(aq)}$$

The oceans tend to mix much more slowly than the atmosphere resulting in large horizontal and vertical changes in CO_2 concentration. In general, tropical waters release CO_2 to the atmosphere, whereas high-latitude oceans take up CO_2 from the atmosphere. CO_2 is also about 10 percent higher in the deep ocean than at the surface. The two basic mechanisms that control the distribution of carbon in the oceans are referred to as the solubility pump and the biological pump.

The solubility pump is driven by two principal factors. First, more than twice as much CO_2 can dissolve into cold polar waters than in the warm equatorial waters. As major ocean currents (e.g the Gulf Stream) move waters from the tropics to the poles, they are cooled and can take up more CO_2 from the atmosphere. Second, the high latitude zones are also places where deep waters are formed. As the waters are cooled, they become denser and sink into the ocean's interior, taking with them the CO_2 accumulated at the surface.

Another process that moves CO_2 away from the surface ocean is called the biological pump. To enable growth of marine plants such as phytoplankton, absorption of CO_2 and other chemicals from sea water are necessary to make plant tissue. Microscopic marine animals, called zooplankton, eat the phytoplankton and provide the basis for the food web for all animal life in the sea. Because photosynthesis requires light, phytoplankton will only grow in the near-surface ocean, where sufficient light can penetrate.

Although most of the CO_2 taken up by phytoplankton is recycled near the surface, a substantial fraction, around 30 percent, sinks into the deeper water before being converted back into CO_2 by marine bacteria. Only about 0.1 percent reaches the seafloor however, to be buried and form part of the sediments. However, almost no phytoplankton seem to grow faster in higher CO_2 environments, unlike many land plants. This is because phytoplankton growth in the oceans is generally limited by the availability of light and chemicals other than CO_2, principally nitrogen and phosphorus but also smaller amounts of iron, zinc, and other micronutrients.

One proposed approach for enhancing carbon removal from the atmosphere is to enhance phytoplankton growth by fertilizing specific regions of the ocean with a relatively inexpensive biologically limiting chemical like iron. The hypothesis is that the resulting bloom of oceanic plants would remove CO_2 from the atmosphere then transport that carbon into the deep ocean or sediments, effectively removing it. The effectiveness of the "iron hypothesis" is being tested with several research efforts attempting to scale up iron fertilization experiments. But it is doubtful that sufficient increases in CO_2 will reach the sediments of the ocean floor, but rather be taken up by other marine life. Other carbon sequestration approaches such as direct injection of liquefied CO_2 into the deep ocean are also being examined. Expense is likely to be a limiting factor with this proposal.

Two billion years ago, bacteria evolved that were capable of consuming atmospheric carbon dioxide to produce energy while releasing oxygen gas as a byproduct in a process known as *photosynthesis. These photosynthetic bacteria, or cyanobacteria, were responsible for oxygenating the atmosphere of early Earth, which paved the way for life on land. Researchers recently discovered at least one type of cyanobacteria that can "cheat" at the game of photosynthesis in order to conserve scarce nutrients. The cyanobacterium Synechococcus uses a form of photosynthesis that does not require carbon dioxide to produce energy for the cell. The researchers were studying Synechococcus in order to understand how these cyanobacteria can thrive in parts of the ocean where access to iron, a mineral typically necessary for performing photosynthesis, is limited.*

The bacteria bypass the stage of photosynthesis that requires iron, which in turn prevents the process from reaching the point at which carbon dioxide is turned into stored energy.

It appears then that some organisms short-circuit the complicated process of photosynthesis and instead use light in a minimalist way to power cellular processes directly. Measuring the effect of cyanobacteria that do not produce carbon storage molecules will be essential in determining the net production of carbon compounds from carbon

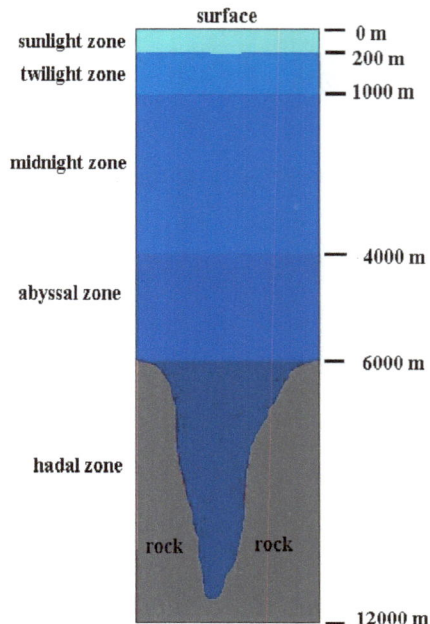

187

dioxide in open ocean ecosystems.

The Ocean Depths

From the surface to the deepest depths, oceans are home to a diversity of life . There are large creatures and microscopic ones; long and short, the multicoloured and the drab, those that just sit while others never stop swimming. There are even some organisms that light up.

Oceanographers divide the ocean into five broad zones according to how far down sunlight penetrates:

- *the epipelagic, or sunlit, zone: the top layer of the ocean where enough sunlight penetrates for plants to carry on photosynthesis.*
- *the mesopelagic, or twilight, zone: a dim zone where some light penetrates, but not enough for plants to grow.*
- *the bathypelagic, or midnight, zone: the deep ocean layer where no light penetrates.*
- *the abyssal zone: the pitch-black bottom layer of the ocean; the water here is almost freezing and its pressure is immense.*
- *the hadal zone: the waters found in the ocean's deepest trenches.*

Plants are found only in the sunlit zone where there is enough light for photosynthesis, however, animals are found at all depths of the oceans though their numbers are greater near the surface where food is plentiful. Even under the Antarctic ice we find an abundance of life. Amazingly, over 90 percent of all species dwell on the ocean floor where a single rock can be the domain of over ten major groups such as corals, mollusks and sponges.

Above is a representation of the various ocean zones, from the warm sunlit waters at the surface to the cold dark depths of the Mariana Trench, the deepest point in the oceans at 11,033 m. Notice how shallow the sunlit zone is compared to the midnight or abyssal zones.

There are two general types of plants found in the ocean, those having roots that are attached to the ocean floor and those not having roots which simply float or drift about with the current. The rooted plants in the ocean are generally found in shallow water because there is not enough sunlight to sustain photosynthesis in deeper waters. Since sunlight does not penetrate more than two hundred metres into the ocean, most of the ocean is not capable of supporting rooted plants. Nevertheless, plants are found throughout most of the oceanic surface waters.

The most abundant plants in the ocean are known as phytoplankton. These are usually single-celled, minute floating plants that drift throughout the surface waters of the ocean. A bucket of sea water might hold a million microscopic diatoms which are relatives of seaweed encased in clear epidermis. In order to grow, phytoplankton need nutrients from the sea water and lots of sunlight. The higher light density occurs in the tropics but nutrients there, especially nitrogen and phosphorus, are often in short supply. When large quantities of diatoms and other phytoplankton are present they give a colour to the

189

sea. Spectacular phytoplankton blooms are found in cooler waters where nutrients are brought up from the sea floor during storms.

Life finds a way:

Having said this, recently researchers have made the discovery that green sulphur bacteria living near hydrothermal vents has major implications for where photosynthesis happens and where life may reside. A team has found evidence of photosynthesis taking place deep within the Pacific Ocean. They found a bacterium that is the first photosynthetic organism that doesn't live off sunlight but from the dim light coming from hydrothermal vents nearly 2,400 meters deep in the ocean.

The discovery of the green sulphur bacteria living near hydrothermal vents off the coast of Mexico has significant implications for the resiliency of life on Earth and possibly on other planets, said the research team.

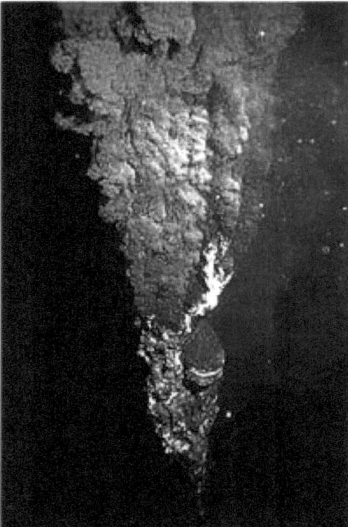

Thus we concede that the ocean still holds many mysteries as well as clues to improve our knowledge on the beginnings of life on this planet.

The bacteria apparently live in the razor-thin interface between the extremely hot water (350 degrees Celsius) coming from a flange vent and the very cold water (2 degrees

190

Celsius) surrounding it. The research team was from the University of British Columbia and published their discovery in an article titled "An obligately photosynthetic bacterial anaerobe from a deep sea hydrothermal vent," in the Proceedings of the National Academy of Sciences (June 2007). On applying DNA analysis, the team classified the microbe as a member of the green sulphur bacteria family, which use light and sulphur to obtain energy. The organism relies solely on photosynthesis to live which is startling in the sense that one does not expect to find photosynthesis in a place that is so completely dark.

These organisms are the champions of low-light photosynthesis, having the most elaborate and sophisticated antenna system. The antenna system of the bacteria uses a chlorosome complex, which basically acts like a microscopic satellite dish, to efficiently collect any light it can and transfer it to the organism's reaction centre where the photosynthesis takes place.

Photosynthesis is something that is not limited only to the very surface of our planet. It conjures up the possibilities where one might find photosynthesis on Earth, as well as on other planets. Europa, a planet-sized satellite of Jupiter is far too distant from the Sun for traditional forms of photosynthesis. It is believed that under the ice-covered surface of Europa are liquid oceans. At the bottom of those oceans it is speculated there might be very hot thermal vents. Those vents could harbour the potential for spawning photosynthetic organisms.

This is just one example of life enduring harsh and extreme environments and is some evidence that life forms may have survived travelling across the cosmos to arrive at our Earth in its youth.

Overfishing

We have seen on the news many valiant attempts by Greenpeace to bring an end to Japanese whaling. Whale meat was a delicacy for centuries in Japanese cuisine. After signing an agreement in the 1980's to ban the practice, Japanese whaling ships continued to hunt whales over the next twenty years on the pretext that it was purely for scientific research. Under the treaty set up by the International Whaling Commission, whose raison d'être is to safeguard whale stocks across the world's oceans, only certain indigenous groups are permitted to practice the hunting and killing of whales for food. Factory ships don't get a mention!

But whales have been prized in Japan as a source of both food and a variety of byproducts, and Japanese whalers caught 2,769 whales in 1986. Japan ended commercial whaling in 1987, following the imposition of a worldwide ban on the hunting of endangered species of whales by the International Whaling Commission. It announced that it would catch 875 whales for "research" purposes. In the year 2000, the Japanese whale catch of over 16,700 toothed whales of various species represented about 82% of the world's whale catch.

The biggest single threat to marine ecosystems today is overfishing. Our appetite for fish is exceeding the oceans' ecological limits to satisfy the demand. Already we have seen varieties of once popular table fish no longer available. Ships using state-of-the-art fish-finding sonar can pinpoint schools of fish quickly and accurately. The ships are fitted out as factories, containing fish processing and packing plants, freezing systems, and powerful engines to drag fishing nets and lines through the world's oceans. , such as cod, flounder, halibut, tuna, swordfish, marlin and skate have been fished out since these large scale industrial fishing fleets began in the 1950s. The depletion of these top predator species may cause a shift in entire oceans ecosystems where commercially valuable fish are replaced by smaller, plankton-feeding fish. In the 21st century we already see bumper crops of jellyfish replacing those fish consumed by humans. The cod fishery off Newfoundland, Canada collapsed in 1992, leading to the loss of some 40,000 jobs in the industry. The cod stocks in the North Sea and Baltic Sea are now heading the same way and are close to complete collapse. The problem is akin to having a single large farm but accessed by a phalanx of farmers each competing for his own share. Instead of trying to find a long-term solution to these problems, the fishing industry is turning towards other oceans to fish. There is a continued rejection of the advice provided by scientists about how these fish stocks should be managed and the need to fish in a sustainable way.

An example is the sustainability of Bigeye and Yellowfin tuna stocks in the eastern Pacific Ocean. Populations of both are on the decline, both in overall population and individual fish size. Bluefin and Albacore Tuna may be added to this list.

In the seas off Alaska, modern factory fishing started in the 1960s, when large Japanese and Soviet factory stern-trawlers replaced the smaller, less efficient side-trawlers. Catches of Pacific Ocean Perch, Pacific Herring and Yellowfin Sole reached record levels by the early 1960s, quickly followed by collapses as each stock was overfished. As stocks of one species crashed, the fleets shifted their fishing effort to another species.

Generally speaking, overfishing happens when fishing removes a volume of fish from a fish population in numbers that exceed the stock's ability to replenish itself. Diminishing numbers of fish without a corresponding decrease in the fishing, eventually leads to a population collapse, rendering the fish stock commercially extinct.

Recently, marine scientists have described ecological extinctions of marine megafauna i.e populations of whales, manatees, dugongs, monk seals, sea turtles, swordfish, sharks, giant codfish and rays from overfishing on a global scale. But we must ask "is it just the world fishing industry that will suffer or is their likely to be a more hideous and severe effect on the total ocean ecologies?" The thrust of this writing is on increased greenhouse gases. "How can a dying sea assist in such an urgent and pressing problem?" is my additional question!

As aforementioned, Japan is one of the world's foremost fishing nations, accounting on average for about 8.5% of the world's catch. In 2008, its total catch was in excess of 5.5 million tonnes, ranking third *in the world. The estimate of total world tonnage(2008) is around 82 million tonnes annually (compare to 58 million tonnes in 2000).*

Japan has seen conflict with Canada over salmon, with the former USSR over fishing in the Sea of Okhotsk and other Soviet waters , with the Korea and China over their limitations on Japanese fishing operations, with Australia over pearl fishing in the Arafura Sea, with Indonesia over fishing in what Indonesia regards as inland waters, and with the United States, especially over fishing in the north Pacific and Alaskan waters. Japan has been adversely affected by the adoption of the 200 mile fishing zone by the United States and more than 80 other nations. Fishing in waters claimed by the United States (where about 70% of the Japanese catch originates) or by many other nations now requires payment of fees and special intergovernmental or private agreements.

In summary, more than 40% of the world's fishing is carried out unsustainably and largely in defiance of international codes of conduct. It is now thought that voluntary schemes to prevent overfishing should be replaced with binding international laws that can better protect marine ecosystems.

Scientists graded the 53 major fishing nations i.e those that take 96% of the world's marine catch, on how their intentions matched actions in complying with the UN's code (1995). The code sets out criteria on how countries should implement the right type of equipment for how fish are caught and how to minimize ecosystem impacts such as catching unwanted fish species that have to be thrown back into the sea and minimizing effects on dolphins and other mammals.

Norway comes top of the list with a compliance rate of 60%, followed by the United States, Canada, Australia, Iceland and Namibia.

In the bottom 28 countries, and representing more than 40% of the world's marine fish catch, the compliance rates were so poor that they achieve "fail" grades, meaning they complied with less than 40% of the UN code of conduct. Twelve countries in this category also failed in all or most sections of the compliance analysis.

Australia has introduced many marine parks where fishing is limited as well as identifying outmoded and dangerous fishing techniques harmful to endangered species. Poaching by Indonesian and other nations has been a frequent occurrence in Australian waters that are regularly patrolled and protected by various state government departments and the Royal Australian Navy.

Top fishing nations include Japan, European Union, Russia, Peru, Chile, Liberia, Morocco, China and Argentina. However 75% of the world marine fisheries catch (now exceeding over 80 million tonnes per year) is sold on international markets. This presents difficulties to the logic of gathered statistics when citing any particular nation's catch. I am tempted here to draw one of my famous graphs, this time referring to world fish stocks. But why insult my reader's intelligence? It would be too depressing to view!

Ocean Dumping

Ocean dumping is defined as the dumping or placing of a wide range of materials, including garbage, construction and demolition debris, sewage sludge, dredge material, and waste chemicals into the ocean. Ocean dumping may be regulated and controlled in certain cases, while in others, ships and tankers dump haphazardly or illegally within coastal waters. But since the Clean Water Act was passed and reauthorized in the 1970s and 1980s, these activities have been curtailed to a great extent.

Today most of the pollutants in the ocean are caused by day-to-day human activities. These include: release of fossil fuel and waste combustion in the atmosphere, pesticides, toxic-waste products, nutrients, and sediments that enter the water as runoff from the land.

The relative contribution of dumping to the overall input of potential pollutants in the oceans is estimated at 10%, a less than satisfactory situation. Human activities contribute the following sources:

Run-off and land-based discharges (54%)

Land-based discharges through the atmosphere (33%)

Maritime transportation (12%)

Offshore productions (1%)

The waste that we produce from our daily activities reaches the ocean either from direct dumping or as run-offs through drains and rivers and include such things as oil, fertilizers, solid garbage, sewage and toxic chemicals. Oil spills are thought to cause about 12% of ocean pollution. As much as 36% of this originates from the drains and rivers as runoff from cities and industry. In coastal areas fertilizer runoff from farms and lawns are major contributors to ocean pollution Solid garbage, if not disposed of properly, ends up in the sea. Major items are household products such as plastic bags, balloons, glass bottles, shoes, and packaging material. Untreated or undertreated sewage pollute the seas to a large extent particularly in Africa, South America, Asia and Southeast Asia via major rivers. Pesticides and chemicals used in common household products pollute the ocean and affect all forms of marine life, particularly those such as detergents with high phosphate.

Smoke emanating from a cars, trucks and factory smoke stacks ends up combining with atmospheric moisture to form acid rain. Acid rain is pollution such as oxides of nitrogen and sulphur mixed with regular rain. When acid rain falls over the ocean, (and most of it does) it changes the water chemistry and alters the ecosystem of the sunlight zone. Pleasure boating activities result in the boat's engine when it is running giving off excess gasoline, which enters the water. Medical waste or clinical waste refers to biological products which are essentially useless. Disposal of this waste in the ocean causes extreme forms of damage to marine life.

The London Convention (1970-1980) regulated the dumping of waste in ocean waters. Currently there are 81 parties to this Convention. Dumping of high-level radioactive wastes has never been allowed under the London Convention. However there is strong evidence that several nations dumped radioactive materials offshore from the 1940's for several decades. Since 1983 a moratorium on the dumping of low-level radioactive wastes has been in place pending the completion of scientific and technical studies as well as studies on the wider political, legal, economic and social aspects of radioactive waste dumping. The parties agreed in 1993 to ban dumping of all radioactive wastes in the ocean. This legally-binding prohibition entered into force on 20th February 1994. The 1996 Protocol entered into force on March 24, 2006 with the aim of eventually replacing the London Convention. So far, 30 States have acceded to the 1996 Protocol. States can be a party to either the London Convention 1972 or the 1996 Protocol, or both.

But despite the treaties, there are no international policing efforts. The language of the treaties is vague about the enforcement of the treaty. Enforcement is under national, not international control and the vessels are subject to the laws of the country where they are registered. This has resulted in a despicable but common practice called 'flag-of-convenience' under which shipping companies re-flag their vessels in countries that have lower environmental standards. Many countries where ship companies register their flags aren't members of these treaties, so the companies don't have to abide by the standards. This is back to the old chestnut of the rich and smart dumping their waste on the poor and ignorant. And notice here I say 'companies', meaning the international super-powerful global enterprises with dark histories. The responsibility of this illegal dumping is the source company or government organization of the waste i.e the original owners. 'Out of sight out of mind' is not good enough in these critical days and more prosecutions must follow. They should even be rep if the act of dumping has been performed since the signing of the protocol.

Sewage from cruise ships, dumping of bilge waters from container ships and other large vessels have also contributed to ecological disasters both from chemical pollution and introduction of marine species from remote locations. There was a time when the world's human population was small and the oceans could bear all this. But not any more. We either enforce international law strongly now or see the seventh vile fall upon the Earth and suffer the inescapable consequences.

"And the oceans of the deep shall brood no more but lie still, red and sickly with no breathe of life. The Sun and the Moon shall wax red and darkness shall be upon the face of the Earth. Woe be to the sons and daughters of men in those days for they shall find no relief or place of refuge! Their gardens shall bring forth no fruit or sustenance for want of goodly waters and freedom from salt. The air will burn in their lungs and they will gasp for breathe between their cries of anguish and pain."

... but it doesn't have to be this way!

Twelve   **The Water Car**

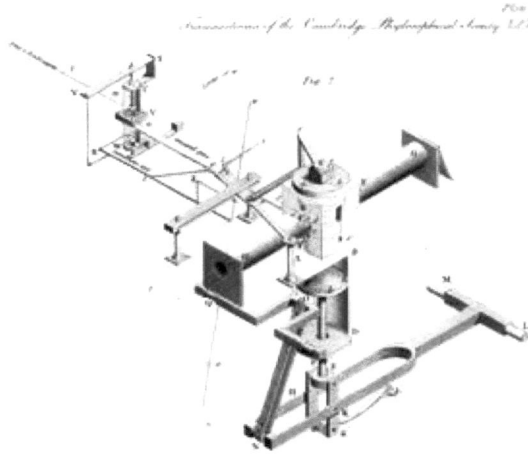

I will commence by referring to the great age of steam. Here water was used to drive machines (including the Titanic), factories, mines and the railways. Those gargantuan engines of iron choofing and hissing, producing sufficient power to launch industrialisation on a scale not seen before in human history. But as charming as steam was it is important to know that it was fossil fuel, mainly coal, that was the source of the energy required to drive these goliaths of the eighteenth and early nineteenth century.   Can we take up the essential positive aspects of this technology and couple to other technologies where the heat energy is supplied but some other means? Alternatively, why not just electrolyse water to form hydrogen and oxygen then use these gases to turn an engine either to produce mechanical energy directly or produce electrical energy some of which can be stored? This is exactly what some researchers in the automotive industry are trying to perfect.

If we go back almost 200 years we find that in 1820, the Rev. W. Cecil wrote a paper entitled: "On the application of hydrogen gas to produce a moving power in machinery; with a description of an engine which is moved by pressure of the atmosphere upon a vacuum caused by explosions of hydrogen gas and atmospheric air." In his document Cecil explains how to use the energy of hydrogen to power an engine and how the hydrogen engine could be built. Here is a sketch and sample of his notes presented to the Cambridge Philosophical Society.

XIV. *On the Application of Hydrogen Gas to produce a moving Power in Machinery; with a Description of an Engine which is moved by the Pressure of the Atmosphere upon a Vacuum caused by Explosions of Hydrogen Gas and Atmospheric Air.*

By the Rev. W. CECIL, M. A.

FELLOW OF MAGDALEN COLLEGE,
AND OF THE CAMBRIDGE PHILOSOPHICAL SOCIETY.

[Read *Nov.* 27, 1820.]

THERE is scarcely any uniform operation in the Arts which might not be performed with advantage by machinery, if convenient and economical methods could be found for setting such machinery in motion. The extensive application of machinery, therefore, depends much upon the number and various capabilities of the engines which can be employed to produce moving force. Even the most perfect engines at present employed for this purpose, are not capable of being applied universally; but each has a province peculiar to itself, beyond which the use of it cannot be extended with profit or convenience.

Two of the principal moving forces employed in the Arts are Water and Steam. Water has the singular advantage, that it can be made to act at any moment of time without preparation; but can be used only where it is naturally abundant.

Notes from CRF Research Laboratory on the Hydrogen Engine:

Hydrogen fuelled internal combustion engines are a potential option and stepping stone to hydrogen fuel cell vehicles. With near zero emissions and efficiencies exceeding port-fuel-injected engines, prototypes have already been demonstrated. In addition, hydrogen powered vehicles can potentially use the existing manufacturing infrastructure for production, a big plus in their favour. Current efforts are focused on developing an advanced spark-ignited engine with efficiencies approaching that of a high-efficiency diesel engine with equivalent power and emissions that are effectively zero. Direct-injection is one of the most attractive options since it has the potential to avoid many problems associated with the use of hydrogen in premixed hydrogen engines, such as pre-ignition and back-flash. In addition, in comparison to a premixed this method avoids the power loss associated with the displacement of air by lighter hydrogen

because fuel is injected after the intake valve has closed. For comparison, a hydrogen engine can deliver approximately 115% the power of a gasoline engine of equivalent characteristics. The challenge now is that in-cylinder injection requires H_2/air mixing in a very short time (approximately 5 ms at 5000 rpm). Incomplete mixing can produce misfire, high

NOx emissions, reduced efficiency, and power loss.

The CRF laboratory houses an automotive-sized single-cylinder engine (~0.6 litres/cylinder) that provides extensive optical access for application of advanced laser-based optical diagnostics to study fundamental in-cylinder engine phenomena. The engine head is a pent-roof, four valve, centre-spark, and side-injection type. Hydrogen injection is through a high-pressure (max. 200 bar) gaseous injector.

New Designs:

These include a rotary engine and the cross "Bean Machine" which cycles hydrogen and oxygen with either water vapour or an inert gas in a completely closed system, and operated at around 200 ^{0}C. Details are sketchy but the principle is a closed system which produces electric current by electromagnetic induction. Oxygen and hydrogen are reconstituted by electrolysis by external solar panelling and direct

PRINCIPAL OF WATER ENGINE

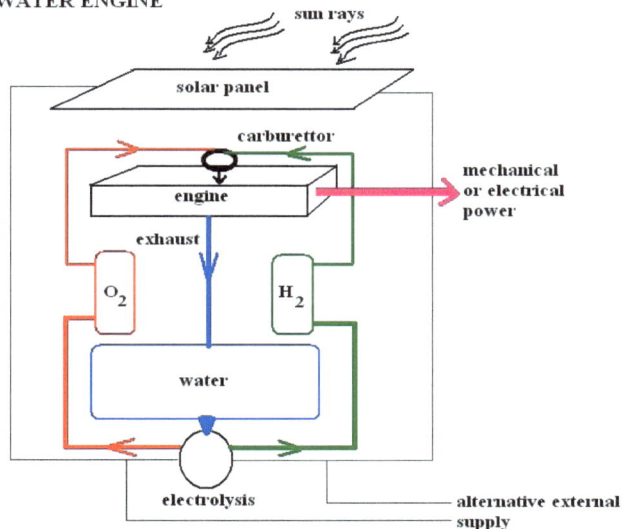

204

charging from a power source. Ideally, new coatings permit just about external surface of the vehicle (windows included) allow for solar charging. The steam or inert gas is essential to prevent damage to engine components due to excessive heat burst during firing. I give here again the brief schematic of the water engine without detail of the detail of the engine itself other than that it may either produce electrical power or mechanical power directly. The idealised system is completely enclosed so that it doesn't even have steam as an exhaust. It has no exhaust at all.

Hydrogen Engine Center Announces a 9.3L Compacted Graphite Iron Engine (June 2008):

Hydrogen Engine Center, Inc. announced that it has entered into an agreement with Eliminator Performance Products to produce the largest spark ignited hydrogen V8 engine yet built. It is intended for large hydrogen-fuelled electrical power generation systems and for buses. HEC's founder, Ted Hollinger, said that the 572 cubic inch engine will provide a much needed power source. Hydrogen is a very light gas and it takes a lot of displacement for every kW of power produced. Compacted Graphite Iron will increase the strength and life of the engine by more than five times and thus give very long engine life which is essential for engines running continuously. This is the company's first Distributed Generation engine. After years of work it looks like HEC has created

an engine that can achieve the efficiency and durability that the industry has long been looking for. The engine is to be commercially built in the United States." Hydrogen Engine Center, Inc. (HEC) develops systems and processes used in the design, manufacture and distribution of alternative fuel internal combustion engines, engine controls and power generator systems. These technologies are for use by customers and partners in the industrial and power generation markets. These solutions and the engines using them are designed to run on hydrogen, ethanol, methanol, ammonia and traditional fuels. Engines and engine products are sold under the brand name Oxx Power®.

All this is heartening stuff, but it must again be remembered that if an internal combustion engine draws in air, hydrogen internal combustion engines also produce nitrogen oxides (NOx), which contribute to global climate change.

Electric Cars and Hybrids:

We see that many of the major automotive producers such as Honda, BMW, Ford etc. have prototypes and even production cars that are hybrids between electric and petrol driven. Tesla is an American auto company that has produced a commercial electric car with coal face fuel cell technology. It is promoted as a "Green Car" due to its quietness and zero emissions. But can we honestly assign such a lofty tag. It has excellent performance specs and can compete with a modern petrol car. But there remains the one nagging doubt... it must be constantly recharged by plugging in to the conventional electric grid. This electricity is predominantly being produced from fossil fuel consumption, albeit somewhere else and far away!

Another concern is that the high tech batteries also contain not so friendly elements, chemically speaking, and will have to be manufactured as well as dealt with at the end of their life and usefulness. So every silver lining has a cloud hanging not to far above it when we speak of super technology. When presented with the glossy brochure of the new super technology we must question whether this is genuinely an improvement and environmentally more friendly. Without doubt all the innovation and application of modern materials produces a tantalising (but expensive) product. But are their simpler pathways and choices that better guarantee a cleaner future? Its not what the salesperson actually says and promotes that we must be concerned about, it is those essential details she conveniently omits.

Rotary Hydrogen Engine From New Zealand- A Steam Engine Without a Boiler

As hydrogen burns much hotter than our present fuels, the team are looking to turn this into an advantage rather than a problem (as it has been considered in ordinary internal combustion engines, causing an overheating of valves leading to premature ware). What is proposed with the rotary engine is to use this higher temperature of the hydrogen as a catalyst or trigger. Using water injection along with a small amount of hydrogen, on igniting the heat turns the water vapour into high pressure steam on each power stroke. Power is not being sought from the hydrogen alone, it will be derived from the steam with the hydrogen as the internal heat source. This idea is the reverse of conventional steam engines, where a large body of water is heated externally. Rather, just a small body of water is applied internally.

207

To dissociate hydrogen and oxygen from water without electrolysis requires a temperature of 425 degrees C. The water volume will expand 1700 times during that change, exerting a most tremendous force. This force is elastic, not like the hammer blows of the internal combustion engine. In comparison, the diesel engine obtains its heat from high compression of the air reaching temperatures of 550 degrees C or more. This though, creates back pressure which results in power loss during the up stroke. With hydrogen and water this is eliminated. The term "stroke" cannot be used here as technically speaking the engine has no piston, a round rotor revolves in a circle in one direction.

The team's aim is to move into the marine field, looking to get away from the burning of fossil fuels and its consequential pollution of the planet. In any case oil resources are limited and naturally will one day be gone.

BMW has succeeded in developing a monovalent hydrogen engine with a diesel-typical geometry and progressive hydrogen high-pressure direct injection technology. Efficiencies on par with those of the best turbo diesel engines with a maximum of up to 42 percent, as well as delivering a power output of 100 kilowatts (about 134 hp) per litre of engine displacement have been achieved. The newly developed combustion system combines spark-ignition and diesel combustion technology to better exploit the favourable burning properties of hydrogen. The experimental system also involves direct-injection technology, supplying high-pressure injectors that can push the hydrogen into the chamber at pressures of up to 300 bar.

It should be noted here that the storage of enough hydrogen on-board to allow cars to travel long distances is a major technical hurdle for hydrogen vehicles. But as more efficient engines are created, in future less hydrogen will be needed to be used and stored. Some observers are still wary on the panacea of hydrogen as a fuel for cars; hydrogen fuel is much more expensive than fossil fuels and there is no infrastructure, such as hydrogen service stations, to support a large-scale roll-out of hydrogen powered vehicles. While hydrogen is the most abundant element in the universe, capturing it, distributing it and storing it is complex and costly. Hydrogen should be the ultimate future fuel, but it can't solve the immediate problems of greenhouse emissions and oil shortages. "Hydrogen was 40 years away 40 years ago. It's still 40 years away" said David Lamb, Low Emissions Transport Leader at CSIRO, Newcastle Australia. Fortunately not all the world shares his opinion!

Tesla is currently taking orders for their Roadster Sport, an exceedingly high performance sports car based on the all-electric zero emission world leader, the Roadster. Specifications include:

Motor: Hand wound stator and high density windings for lower electrical resistance and high peak torque; 0-100 kmph in a stunning 3.7 seconds!

The Tesla Model S electric car, base price $US50000, is a sleek four-door with a range of up to 480 kilometres on a single charge. Tesla hopes to build 20,000 of the sedans per year by mid-2012. Hiding somewhere below and supported by a Lotus styled chassis is a 440 kg battery pack consisting of more than a handful of 3.7 volt lithium based cells.

It was said that the Model X was surely the most-awaited electric vehicle of 2015. Perhaps the most-awaited EV of all time... though, the broader accessibility *of the Model 3 may have it beat. It is a super-high-performance, utilitarian, luxury SUV or crossover with falcon-wing doors.*

The Rimac Concept_One is not everyman's car. It is an electric supercar out of Croatia that costs a fortune… as in, $1 million.

The Chinese company BYD Auto has a vehicle called the BYD F3DM. This Dual Mode (range-extended) electric vehicle hasn't sold in numbers that would challenge comparable American counterparts like the Chevy Volt. It has been reported that since its December 2008 launch, only 80 units have sold over the first couple of months (20 being to China's government). High price, concerns how long the battery will last and reliability issues might be to blame. But then the Tesla hasn't sold that many cars to date as we have seen above. Currently the new electric car costs around $US22000 and is available only to Chinese fleet buyers. If a future deal to replace China's taxies that would make a huge difference. As it stands, the F3DM is still looking for market share.

BYD's gasoline-only counterpart to the F3DM is the F3, and is available to the Chinese public. In March 2009, reports that 20,940 F3 cars sold.

The 2015 BYD Tang:

The Tang looks even more attractive and I think will help to mainstream electric vehicles in China. It is another plug-in hybrid, and accelerates to 60 mph in 4.9 seconds. BYD is hoping to sell thousands of Tangs a month, a feat few electric vehicles have reached.

The BYD company plans to lower the price further to around $US16100 once they can begin mass production. That would likely convince buyers that the company is prepared for a long life cycle with the vehicle. The number of available charging stations for electric vehicles in China is another issue. Independent studies and time will soon discern what has been said about battery life and vehicle reliability.

Whereas electric cars will reduce dependency on gasoline, the production of electricity in China still relies heavily on coal fired power stations, the number of which is increasing at an alarming rate. It is to be hoped that Chinese companies and Government research institutions will also seize the moment to develop the hydrogen car as well. It alone can be the true saviour of the environment!

Epilogue

What can I say? Probably the most important aspect of this writing is to again emphasise that the turn around in social behaviour in order to halt global warming, climate change and an eventual catastrophic nemesis of society as we currently know it lies in the hands of ordinary people. Big Government is too closely related to and intertwined with Big Business. It is at the level of local government that the voice of the people may be heard and real action taken. This has already been recognised and demonstrated where local communities have taken decisions to forge a new path with alternative energy, the reduction of waste, the greening of their immediate environ and a collective awareness on the necessity of creating new ways to achieve these. Education of the young is paramount to a solution.

On the question of exponential population rise, each nation must look fifty and one hundred years ahead and plan its own future to halt this madness and/or make adequate preparation to maintain a safe and healthy society as part of the global community. Finger-pointing and abuse of one's neighbour is unhelpful compared to assistance in terms of ideas and technology backed up with financial support and educational stimulus. The greed of the West must be tempered so that poorer nations do not bear economic and environmental hardship causing a general degradation in their every day life requirements. We have the knowledge to beat disease, a lack of clean water and hunger which leads to severe infant mortality increase and an infectious malaise in that society which further aggravates these problems.

Whilst heading in the right direction, the weaknesses demonstrated in Kyoto, Bali and Santiago (Chile) has been the uncompromising stance of some nations in their adherence to lengthy time frames for real change, ignoring the urgency and need for prompt action with regard to reforestation, reducing carbon emissions and implementation of alternative energies. "We can reach such and such a target by 2030" and similar! Unfortunately the world has been placed in greater jeopardy with the onslaught of widespread recession, providing an untenable but authentic excuse for delaying those strategies so badly needed to save us from overheating. Paris 2015 relies on individual nations making their decisions despite overwhelming recognition of the coming challenge, no longer a fantasy! Unfortunately the constipated economists will always have their sway in the boardrooms of the energy giants of the world.

Apart from the Eureka of nuclear fusion energy, many attempts at alternative energies are fraught with high cost and an ever increasing sophistication in technology and materials that bring their own environmental clout to an overworked and fragile system(examples include hybrid cars and some prototypes for fuel cellsand other high tech batteries). My own stance on Nuclear Fusion as a provider of electrical energy is that, in the long term, nuclear waste will bring an anathema to the human environment and condition that will marginalise even the predicted effects of carbon emissions. No government on the planet has a successful and long term plan (thousands of years) to cope with nuclear waste and hence avoid the hideous health effects that will burden future generations. Dollars earned from Uranium sales easily cloud the mind of economists and politicians as well as an abundance of misinformation and a dangerous mindset of political extremism to protect and maintain a misguided vision of cheap electricity. Nuclear waste will be seen in the distant future as the gilded epitaph of societies of the first half of the twenty-first century. We will be considered as greedy, selfish despots and hedonists not caring for our planet and its future generations. So I plea and appeal to all humans to think about their daily life-style and how each of us can change our bad ways and genuinely strive to improve the Earth, our home in space. We must not be trapped by the glossy magazines of high materialism and the Gods of pleasure and overindulgence. Gluttony is a deadly sin. A blind eye to our poorer neighbour unforgiveable. Remaining silent when you observe someone or an organisation deliberately despoil the land for some short term gain an equally despicable act as that of the despoiler! The extravagance of war and weaponry instead of national investment in parklands and wilderness, roads, houses, schools and hospitals are the hallmark of irresponsible government. In the overwhelming majority of nations, due to systematic building of political institutions and frameworks, particularly on the party basis, nearly always the wrong people are promoted for election to the controlling seats of power. I don't have a solution to change this but it is self evident. Money together with social position and power (often inherited) is the ladder of corruption whence these brokers scale to the topmost echelons of the political scene. It is a sad indictment when there are so many good men and women so much more suited to the position. And when a good man or woman does make it occasionally, there are hideous mechanisms in place to squeeze, denigrate and humiliate with the result that they are forced to step aside for the status quo. Neither am I suggesting here that some extreme of politics is the answer. We have seen enough of communism and other forms of fascism and the consequences throughout the twentieth century. The toll on human life in that century was perverse to say the least.

Too much of a nations best in terms of people, resources and righteous direction of activity have been squandered resulting in intellectual, physical and economic poverty of its people. It appears easier to create a wall of arguments and reason why not to go down the sensible and correct path for an energetic, creative and healthy society. 'Laissez Faire' and 'Business as Usual' are the enemies of us all. Time to awaken and take stock of our current position and embrace the challenges, crashing through the pins and columns of intransigence for a future planet Earth the way it ought to be and how we wish it to be.

Changzhou, China May 2009.
Tongio, Australia December 2015

Appendix A   Small Molecules and some Chemical Reactions

$O_3 + O \rightarrow 2O_2$

Ozone + atomic oxygen → oxygen

$2NO + 2CO \rightarrow N_2 + CO_2$

Nitric oxide + carbon monoxide → carbon dioxide

$2CO + O_2 \rightarrow 2CO_2$

Carbon monoxide + oxygen → carbon dioxide

$C_nH_{2n+2} + (3n+1)/2 \ O_2 \rightarrow nCO_2 + (n+1)H_2O$

Hydrocarbon + oxygen → carbon dioxide + water

$2H_2O_2 \rightarrow 2H_2O + O_2$

Hydrogen peroxide → water + oxygen

$CO_2 + 2H_2 \rightarrow CH_3OH$

Carbon dioxide + hydrogen → methanol

$C_6H_{12}O_6 \rightarrow 2CH_3CH_2OH + 2CO_2$

Glucose → ethanol + carbon dioxide

$CH_3CH_2OH \rightarrow CH_3COOH + H_2O$

Ethanol → acetic acid + water

$CH_2CH_2 + H_2 \rightarrow CH_3CH_3$

Ethane + hydrogen → ethane

$2H_2O \rightarrow 2H_2 + O_2$

Water → hydrogen + oxygen

$6CO_2 + 6H_2O \rightarrow C_6H_{12}O_6 + 6H_2O$

Carbon dioxide + water → glucose + water

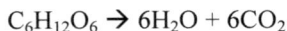

$C_6H_{12}O_6 \rightarrow 6H_2O + 6CO_2$

Glucose → water + carbon dioxide

$N_2 + O_2 \rightarrow 2NO$

Nitrogen + oxygen → nitric oxide

$2NO + O_2 \rightarrow 2NO_2$

Nitric oxide + oxygen → nitrogen dioxide

$NO_2 + H_2O \rightarrow HNO_2 + HNO_3$

Nitrogen dioxide + water → nitrous acid + nitric acid

$S + O_2 \rightarrow SO_2$

Sulphur + oxygen → sulphur dioxide

$2SO_2 + O_2 \rightarrow 2SO_3$

Sulphur dioxide + oxygen → sulphur trioxide

$SO_2 + H_2O \rightarrow H_2SO_3$

Sulphur dioxide + water → sulphurous acid

$SO_3 + H_2O \rightarrow H_2SO_4$

Sulphur trioxide + water → sulphuric acid

$N_2 + 3H_2 \rightarrow 2NH_3$

Nitrogen + hydrogen → ammonia

$FeO.Fe_2O_3 + 2C \rightarrow 3Fe + 2CO_2$

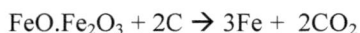

Iron oxide + carbon → iron + carbon dioxide

$CaCO_3 \rightarrow CaO + CO_2$

Calcium carbonate → calcium oxide + water

Appendix B   Hydrogen Fusion and the Creation of the Elements

This is the nuclear fusion process which fuels the Sun and other stars which have core temperatures around 15 million degrees Kelvin*.

[* Degrees Kelvin = degrees Celsius + 273]

*Consider the **nuclear fusion reaction**:*

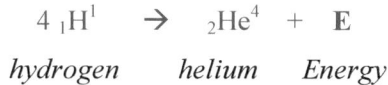

$$4\ _{1}\text{H}^{1} \quad \rightarrow \quad _{2}\text{He}^{4} \ + \ \textbf{E}$$

hydrogen   helium   Energy

Fusion of a single kilogram of hydrogen into helium will release around 10^{14} Joule of heat energy, which is over four million times the energy released by burning the same amount of hydrogen with oxygen. Hydrogen fusion in low-mass stars occurs by the proton-proton chain reaction. In the very hot, dense conditions of the star core, all matter is ionised. Hydrogen nuclei, which may be described simply as protons, will have great speed and kinetic energy. They have no electrons bound to them.

Now four protons will rarely collide simultaneously- the probability of such a reaction is low. Instead, the fusion of hydrogen nuclei into helium nuclei occurs in steps of two-particle collisions. In low-mass stars similar to our sun the dominant process for producing fusion is this proton-proton chain reaction. The first step consists of fusing two protons together to form deuterium:

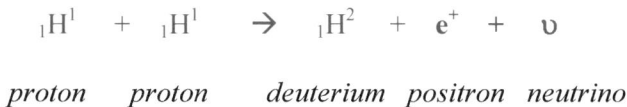

$$_{1}\text{H}^{1} \ + \ _{1}\text{H}^{1} \quad \rightarrow \quad _{1}\text{H}^{2} \ + \ \textbf{e}^{+} \ + \ \upsilon$$

proton   proton   deuterium positron neutrino

In the above equation, $_{1}\text{H}^{1}$ is a nucleus of ordinary hydrogen, which contains only a proton, $_{1}\text{H}^{2}$ is a nucleus of heavy hydrogen called *deuterium which contains a proton and a neutron. The* e+ *is called a positron and is a positively charged electron. A neutrino* υ *is a subatomic particle with no electric charge and a negligibly small mass.*

The next step in the proton-proton chain reaction involves fusing a proton and a heavy hydrogen nucleus:

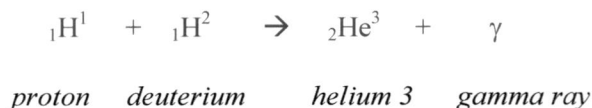

$$_1H^1 \;+\; _1H^2 \;\rightarrow\; _2He^3 \;+\; \gamma$$

   proton *deuterium*   *helium 3* *gamma ray*

In the above equation $_2He^3$ *is a nucleus of light helium containing two protons and a single neutron. The gamma ray is a form of electromagnetic energy.*

Repeating the two reactions above provides two light helium nuclei. The last step involves fusion of these two light helium nuclei to form regular helium and two protons:

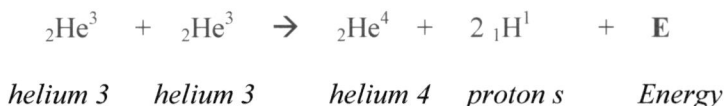

$$_2He^3 \;+\; _2He^3 \;\rightarrow\; _2He^4 \;+\; 2\,_1H^1 \;+\; \mathbf{E}$$

  helium 3 *helium 3*   *helium 4* *proton s*   *Energy*

Thus the overall reaction is the same as that at the start:

$$4 \, _1H^1 \quad \rightarrow \quad _2He^4 \; + \; \mathbf{E}$$

hydrogen   *helium*   *Energy*

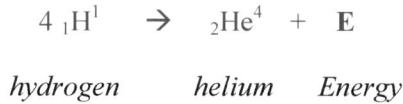

To replicate the conditions of the core of a star is no mean feat. Nuclear scientists have been trying to engineer this since the late 1950's with little success for a safe and continuous system. But the race is on with huge sums of money pouring in to various nations' research programs to achieve what might appear to be the unachievable! It would certainly be the Golden Fleece of clean and abundant energy production, changing the very nature of human civilisation in an unprecedented way.

Synthesis of the Elements

1. Proton-Proton Fusion

The proton-proton process is the nuclear fusion process which fuels stars the size of our own sun having a core temperature less than 15 million degrees Kelvin. A single reaction cycle yields about 25 MeV of energy.

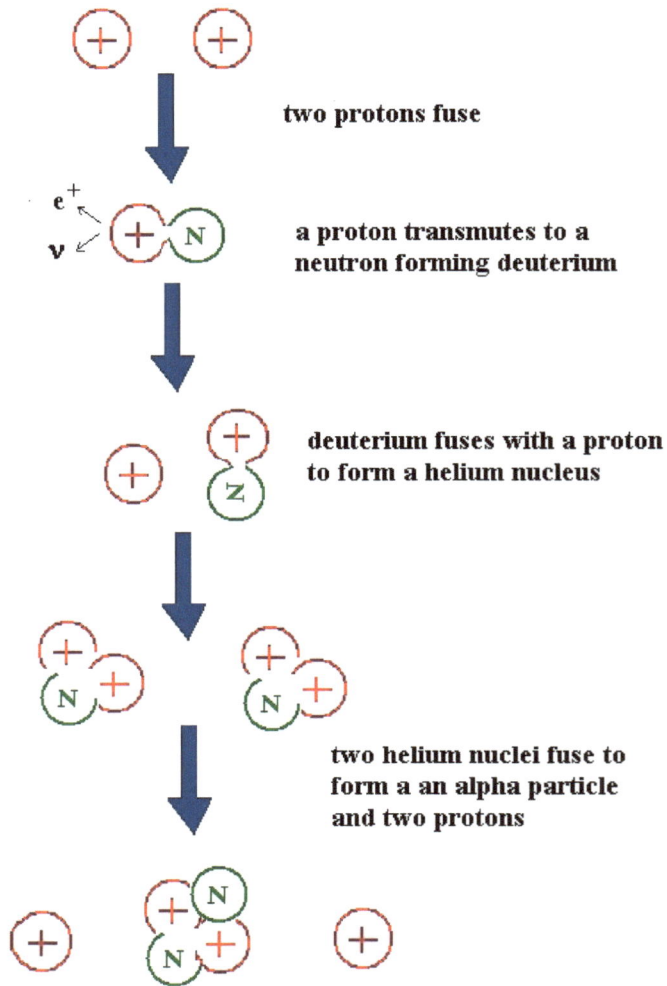

two protons fuse

a proton transmutes to a
neutron forming deuterium

deuterium fuses with a proton
to form a helium nucleus

two helium nuclei fuse to
form a an alpha particle
and two protons

2. The Triple Alpha Process

If the central temperature of a star exceeds 100 million degrees Kelvin as may happen in the later phase of red giants and red supergiants, then helium can fuse to form beryllium 8 and then to carbon 12.

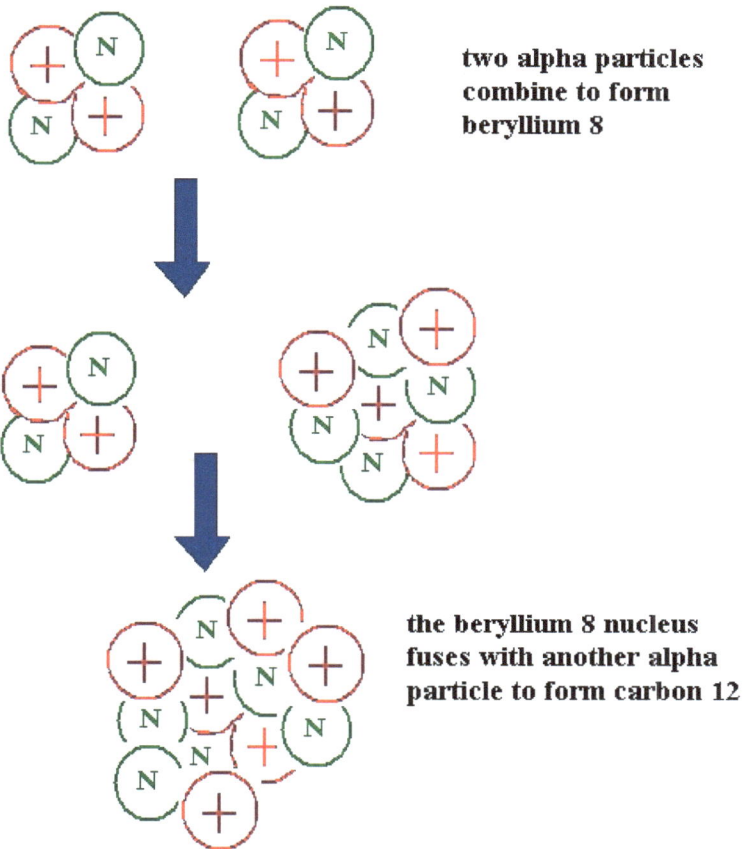

two alpha particles
combine to form
beryllium 8

the beryllium 8 nucleus
fuses with another alpha
particle to form carbon 12

3. The Carbon Fusion Cycle

*In stars with central temperatures greater than 15 million degrees
Kelvin, carbon fusion is thought to take over the dominant role rather
than hydrogen fusion. The main theme of the carbon cycle is the
adding of protons, but after a carbon-12 nucleus fuses with a proton
to form nitrogen-13, one of the protons decays with the emission of a*

positron and a neutrino to form carbon -13. Two more proton captures produce nitrogen-14 and then oxygen-15.

The Carbon Fusion Cycle

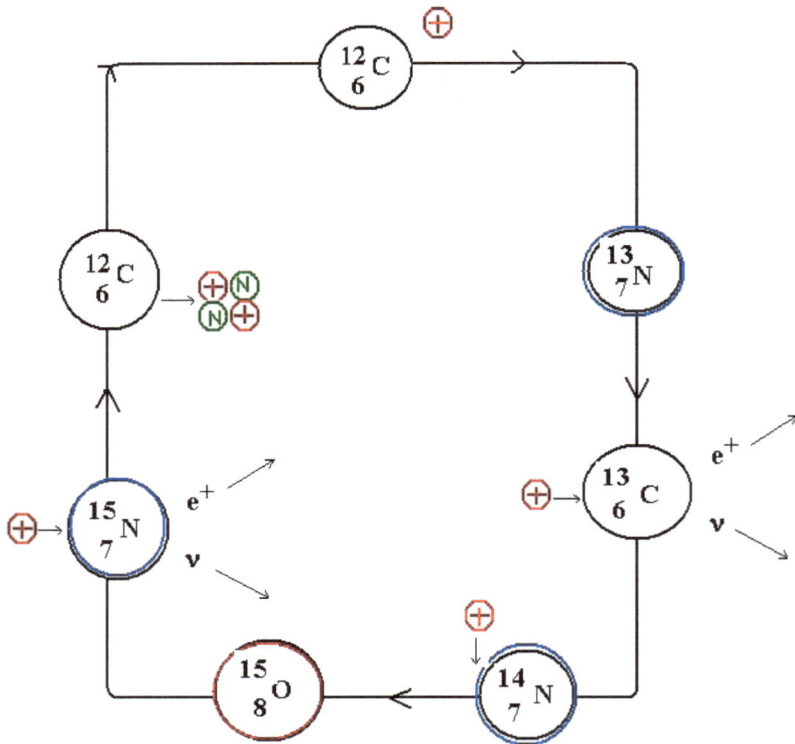

Further neutron decay leaves nitrogen-15. A further proton capture produces oxygen-16 which emits an energetic alpha particle to return to carbon-12 to repeat the cycle. This last reaction is the main source of energy in the cycle for the fuelling of the star.

While this process is not a significant part of the sun's fuel cycle, a star like Sirius with somewhat more than twice the mass of the sun derives almost all of its power from the carbon cycle. The carbon cycle yields 26.72 MeV per helium nucleus.

For his role in working out the energy source for stars more massive than the sun, the carbon cycle, Hans Bethe received the Nobel Prize in 1967. Bethe was one of the outstanding young scientists who fled Nazi Germany in the 1930's

Nuclear Synthesis of the Heavier Elements

In the journal 'Reviews of Modern Physics (1957)' there appeared an article titled "The Synthesis of the Elements in Stars" by E. Margaret Burbidge, Geoffrey R. Burbidge, William Fowler, and Fred Hoyle. In this paper they built a theoretical and computational framework that freshly interpreted forty years of study on the sources of stellar energy and the transmutation of the elements. The problem in nuclear chemistry involves calculating accurate collision cross-sections, which are simply measures of how close one particle must get to another particle before they interact significantly A detailed understanding of collision cross-sections is what enables one to predict nuclear reaction rates and pathways. The smallest of uncertainties in tables of collision cross-sections can lead to erroneous conclusions.

The entire process is known as neutron capture and is responsible for creating many elements that are not otherwise formed by traditional thermonuclear fusion. The remaining elements in nature can be made by a few other means, including slamming high-energy gamma rays into the nuclei of heavy atoms, which then break apart into smaller ones.

Simply put, in the life cycle of a high-mass star, it makes and releases energy, which helps to support the star against gravity. Without this, the big ball of gas would simply collapse under its own weight. A star's core, after having converted its hydrogen supply into helium, will next fuse helium into carbon, then carbon to oxygen, oxygen to neon, and so forth up to iron.

Elements heavier than iron in the periodic table cannot be formed in the normal nuclear fusion processes in stars. For synthesis of elements to iron, fusion yields energy and thus can proceed. But since the iron is at the peak of the binding energy curve, fusion of elements above iron absorb energy. The nuclide ^{62}Ni is the most tightly bound nuclide, but it is not nearly as abundant as ^{56}Fe in the stellar cores, so astrophysical discussion generally centres on iron. In fact an atom of ^{52}Fe can capture an alpha particle to produce ^{56}Ni but this is the last step in the helium capture chain.

Given a neutron flux in a massive star, heavier isotopes can be produced by neutron capture. Isotopes so produced are usually unstable, so there is a dynamic balance which determines whether any net gain in mass number occurs. The probability for isotope creation is usually stated in terms of the cross-section for such a process, and it turns out that there is a sufficient cross-section for neutron capture to create isotopes up to bismuth-209, the heaviest known stable isotope. The production of some other elements like copper, silver, gold, *zirconium and lead is thought to be from this neutron capture process. It is referred to as the "s-process" by astronomers, from slow neutron capture.*

For isotopes heavier than ^{209}Bi, the s-process doesn't seem to work. It is thought that they are formed in the cataclysmic explosions known as supernovae. In the supernova explosion, a large flux of energetic neutrons is produced and nuclei bombarded by these neutrons build up mass one unit at a time to produce the heavy nuclei. This process apparently proceeds very rapidly, in the explosion of the supernova, and is called the "r - process", meaning rapid neutron capture.

To sum up then, atoms heavier than helium up to the iron and nickel may be synthesised in the cores of stars. Those stars of lower mass can only synthesize helium. Stars around the mass of our Sun can synthesize helium, carbon, and oxygen. Massive stars i.e with a mass eight times or more our solar mass, can synthesize helium, carbon, oxygen, neon, magnesium, silicon, sulphur, argon, calcium, titanium, chromium, nickel and iron. Elements heavier than iron are made in supernova explosions from the rapid combination of the abundant neutrons with heavy nuclei.

World Coal Consumption, 1985-2015 (Million Short Tons)

Region	1985	1995	2005	2015	projected 2025
North America	880	1033	1395	1626	1878
Central & South America	28	33	45	53	61
Europe	1580	1117	1223	1425	1645
Eurasia	779	476	507	590	682
Middle East	5	9	19	22	26
Africa	151	175	243	283	327
Asia & Oceania	1465	2273	3764	4386	5065
World Total	4888	5116	7197	8384	9684

World Coal Usage at a Rate of 1.8% p.a Mean Increase in Demand

Year	Demand (pa) million tonnes	World Reserve million tonnes
2000	6000	1000000
2005	6210	987690
2010	6789	954924
2015	7423	919100
2020	8115	879934
2025	8872	837115
2030	9700	790300
2035	10605	739117
2040	11594	683159
2045	12676	621980
2050	13859	555093
2055	15152	481966
2060	16566	402016
2065	18111	314607
2070	19801	219042
2075	21648	114562
2080	23668	334

Appendix D   *Ice Core Sampling of CO$_2$ Levels*

These measurements are directly of CO$_2$ and deuterium (sometimes the isotope oxygen 18) in air bubbles in ice cores. The relationship of deuterium to temperature is an indirect method of calculation which depends on assumptions about relative isotopic abundances in the past. These reconstructed temperatures can only be local, not global. The correlations between CO$_2$ levels with temperature vary from location to location and are not always strong. High resolution of these ice core studies reveal that increase in temperature was actually succeeded by increases in CO$_2$ concentrations, not the other way around! This lag was found to be as much as 2000 years using the deuterium derived temperatures. It is believed that the increase in temperature led to changes in the balance between greenhouse gases in the atmosphere and the concentrations of these gases in the oceans. It is to be noted that ice core temperature series have poor time resolution and cannot reflect any large scale changes which might have occurred in just a few years or decades.

Appendix E   *More Time or no Time at all?   The First Move.*

As mentioned earlier I considered time as a sequence of observable states and that time, space and matter are all connected. What we experience as the passage of time is synonymous with the rate of change of states relative to some adjacent system. This rate of change of states is related to both (a) relative velocity.. Einstein's Special Theory and (b) gravitational field .. Einstein's General Theory. Let us concern ourselves with the latter.

If there is almost no matter present then the gravitational field strength is close to zero. At this extreme, the quantised minimal time interval stretches. Thus, relatively, we see an object or group of objects barely changing state, if not 'frozen'[#].

[# Conversely, if we were an observer in this near zero gravity field area, we would see adjacent systems of high field strength whizzing about rapidly i.e changing states relatively quickly.]

$$\Gamma \rightarrow 0, \; \Delta\tau \rightarrow \alpha$$

At extremely high gravitational field strength the quantised minimal time interval is squeezed so that, relatively, we see a rapid change in states.

$$\Gamma \rightarrow \alpha, \; \Delta\tau \rightarrow 0$$

Thus before matter, there is no time. We might also say that the quantised minimum time interval has stretched to infinity. So in an original state of no space and no matter, all time is less than the blink of an eyelid.

Let me introduce a symbol for pre-existence i.e no matter, no space and no time:

$$N^{\square}$$

like an incomplete N with an empty box at the trailing end. So the universe did not wait eons to be created because there was no time before the point of creation.

Now we all know that in our bank balance it is possible to red figures, zero or black figures. Let's stick with just a dollar. So we may have $1, $0 or $1. I will give them form:

$$O \; , \qquad , \; \bullet$$

233

The geometry of the form is not important. We might describe the first cause as a quantum wave-cum-particle pair formation from nothing... yes nothing. It's akin to setting up two bank accounts with zero balance and then shifting a dollar from one account to another creating a debit in one and a credit in the other, whereas initially I had zero in each. (I was told the American Government do this all the time!) If we say these are newly created particles we see that they must be very different, so different in fact that they produce their own separate space. If they come together they will experience mutual annihilation resulting in nothing again. Personally I don't like this new Universe for a couple of reasons, the main being it is inherently unstable. It is probable that his type of universe has been created over and over many times but is not suitable for the evolutionary path leading to organic structures.

*A more stable structure would be one where its most fundamental particles are created as a trio. In order for annihilation all three particles must come together simultaneously. As this is far less likely than the coming together of two fundamental particles in a two fundamental particle universe, voila! we have arrived at the first move of a stable creation. I am saying then that the first matter was created from **three** particles, each having different properties and initially formed from nothing. Remember what I said before, no space, no time, so there was no waiting time for this to happen, it just happens!*

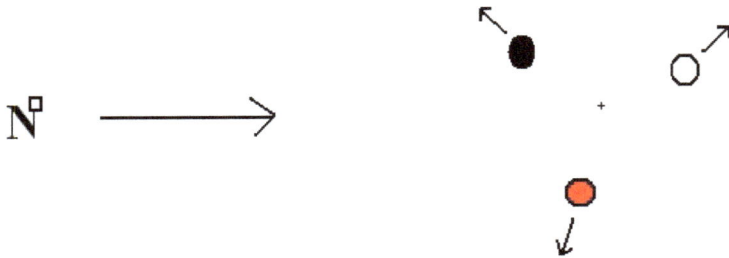

creation of universe from three fundamental particals

The three particles can only disappear or annihilate on condition that all three converge at the same instant. We might view this as three spaces or three coexisting universes each with their own direction of evolution and gravitational field. Naturally we require more than three particles to make a chicken (or egg for that matter).Once the process has started it must avalanche i.e more particles are created in the immediate vicinity and builds to some critical point where we have a sudden expansion which cosmologists call the 'big bang'.

Why this Universe? *We might first ask "how many universes are there, have there been, will there be?" The answer is probably infinite. "Can we ever cross from one universe to another?" The answer is an affirmative "no!" Each universe has its own unique space with no outside so there can be no connection. Even if we were able to 'cross', chances are it would be hostile in the sense that its physical make-up is likely to be different and/or at a different stage in its evolution.*

One Success from many Failures! *I suppose here I am touching upon the anthropological principle. Each universe is self contained, has its own matter, space and time with no boundary. In a nutshell, it is complete in itself. It can never join with another universe but only cycle through its own evolution from creation to annihilation. Not all of these universes, for some reason or other, progress to form stars and galaxies synthesising the heavier elements essential for planetary systems and life. Some are too small, some are too short lived whilst others are not built from the same building blocks as ours.*

So at last we have a success story. A complete self-contained universe that has all the things we need and an evolution of the elements (including carbon) to form stars, galaxies and solar systems with planets. In the far reaches of space containing nebulae of dust from the death of stars, chance reactions eventually form the amino acids. This takes a lot of space and a lot of time but never-the-less has a finite probability. In the depths of the ocean where hot sulphur leaks out we find the most primitive of life forms. This our Universe then has produced living structures that have evolved eventually to produce man. It is as though the universe itself, after coming into being, eventually creates its own consciousness in order to understand itself and what it is. The collective mind of contemporary humans is that consciousness and awareness.

"Are their intelligent beings in other parts of the galaxy Milky Way and in other galaxies?" The probability is a resounding 'yes!' And why? Well because there is uniformity in matter and the laws of science across the universe. This implies that similar paths of

236

evolution have been going on, are going on and will continue to go on upon other worlds with similar attributes to the Earth. There may be dozens in our galaxy. They may be ± 100 Mya from our present state. Whether we will eventually know for certain by communication or meeting is another question. The universe is a big place but the oceans of our planet were once devoid of life but now they teem! And God? Well either the hand of God tickled the void to create the first move or, alternatively, God is the pinnacle of the universe' evolution. But if God is out of the universe, then He is all seeing from start to finish, alpha to omega, genesis to annihilation in the blink of His eye!

Appendix F   ***The First Move Revisited***

It may be that we will never know or understand the first move that brought about the creation of all things. Years ago Fred Hoyle argued that the Universe has always been and that the conundrum of "the chicken and the egg" was mere folly to pursue. He termed this as the Steady State Theory. The proposer of the Big Bang Theory, George Gamow et al also have problems with the detail of the idea of all matter being instantly created from a single point that suddenly came into existence. For one it does not hold with the concept of "energy cannot be created nor destroyed" a basic tenet of physics. For two, the theory follows on from the observation that the universe seems to be expanding and thus originated at some point in the distant past. The expansion might be an illusion or be explained by other causes. The Big Bang, if it is true, might be a continuous process or oscillation of all space and matter where the Universe cycles in its expansion to some maxim and then contracts again to its minim, a single point akin to waves rolling up the beach of a cosmic seashore. The actual creation of all matter and space is just as likely to be occurring all around us as a natural sideshow of our Universe.

We know that hydrogen is both the simplest of all the elements and the most abundant in the universe making up in excess of 99% of all matter. We know that all the heavier elements can be synthesized in the deep interiors of stars and that by nuclear addition of protons and neutrons the heavier elements can be created. But does it follow that all hydrogen was created as a direct result of the Big Bang or can this basic building block also be manufactured from subatomic particles in the vast open stretches of space between the galaxies? If, as I have suggested, the Big Bang is merely part of a cyclical pattern where the slate is wiped clean and space, whilst contracted to a mere fraction of its normal volume, we must look elsewhere for the first move, i.e the creation of something from nothing.

The Primal Pulse

I suggest that the creation of all matter and the Universe as we know must have commenced in the past at a time that is virtually incomprehensible to the human mind. The first microspace was created along with a simple ripple of radiation. This is like an unseen finger plucking at a single violin string, excepting there was no physical string. This vibration, to conform to 'energy cannot be created or destroyed' must have been created alongside

simplest model of the first move

$\Gamma_{A\Omega}$

pre-existence

first move

space creation

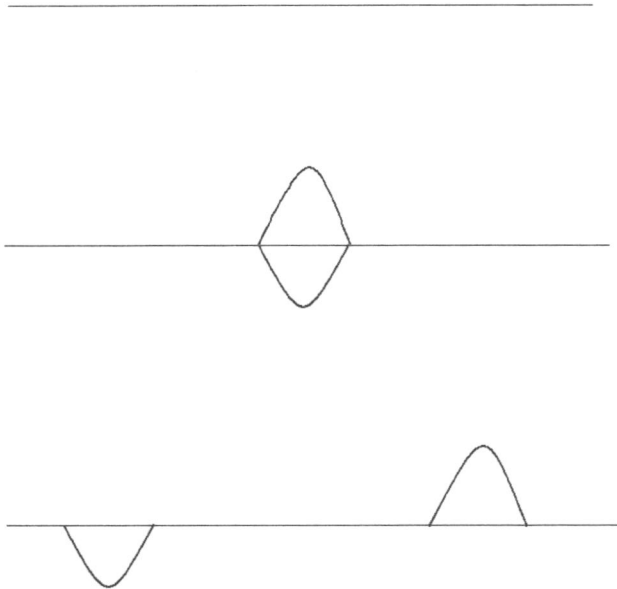

239

its shadow or counter wave so that the net energy remained zero. A first move of this simple pair may not have led to our stable Universe. A creation of three counterpart vibrations whose coincidence or superposition also results in a net energy of zero may have led to our stable Universe in that it is difficult or unlikely that once the umbrella has been opened at the top of the chimney it can never be pulled back in. Due to probability, the chances of annihilation by three components coming together simultaneously and in the exact and necessary special configuration would be rare.

So from a state of pre-existence which I will denote gamma-alpha-omega $\Gamma_{A\Omega}$ the first space and ripple on the pond surface of nothingness appeared. As stated elsewhere, I consider time as quantized, stretchable and dependent

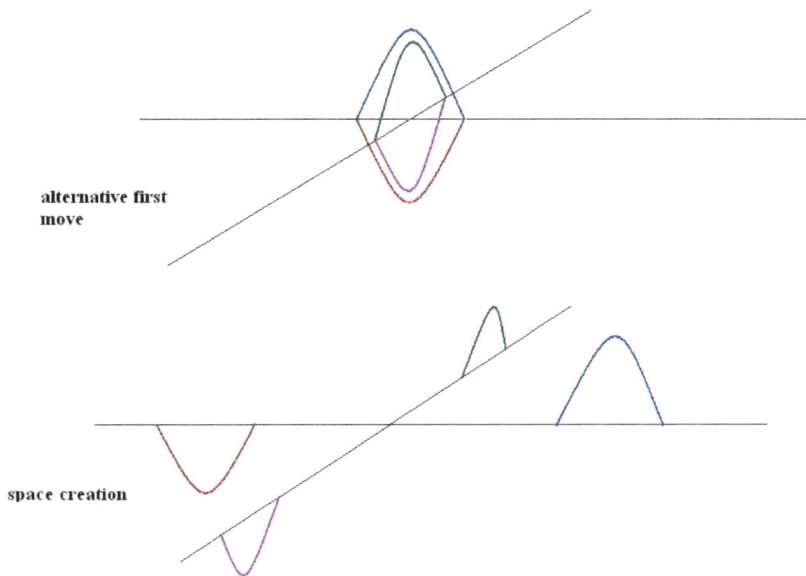

alternative first move

space creation

on space and gravitational field strength. Thus in a condition of nothingness, the time quantum is stretched to infinity i.e no time at all; that is the state of existence of nothingness, $\Gamma_{A\Omega}$, exists for no time*. The trouble with the Big Bang theory is that it poses more questions than answers to the fundamental question of the first move. Why for instance would hydrogen atoms, electrons and the four forces of nature (gravity, electromagnetic,) appear ahead of light? I would have thought that the simplest of particles at the quantum level would have been created first including the photon or electromagnetic wave. In fact I imagine the first move to be the creation of a vibration which taken as a whole sums to nothing. I provide some possible scenarios in the diagrams above. Basically and without reason other than boredom, two halves of a simplistic pulse is created that then move off in different directions.

[* Note: in Appendix E, I used a special symbol $\mathbf{N}^{\square}$ for the idea of "no space"]

If they were superimposed they would again cancel to nothing. I imagine that the first pulse creates its own space and has a resonance in the sense that it encourages the creation of similar pulses in a seeding manner. The deepest question remains… what caused the first blip on the blank screen or why was it formed? As hypothesized elsewhere in these writings, the quantum of time, $\Delta\tau$, in zero gravity (i.e in nothingness) stretches to infinity such that all time of nothingness is reduced to the passage of zero time. So from this perspective one might conclude "it had to happen!"

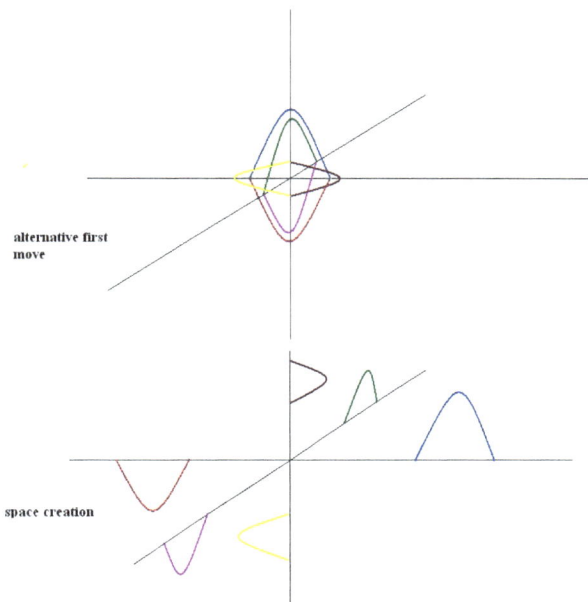

alternative first move

space creation

So electromagnet waves or some precursor waves were seeded from a single first creation of a pair (or higher order set) of primal pulses. Whether we refer these to be waves or particles is unimportant as they will exhibit a dual nature as do other subatomic entities such as electrons and photons. This basic model of the first space and primordial particles-cum-waves may explain why photons travel at the speed of light. In some sense they are space creators.

$\Delta\tau$

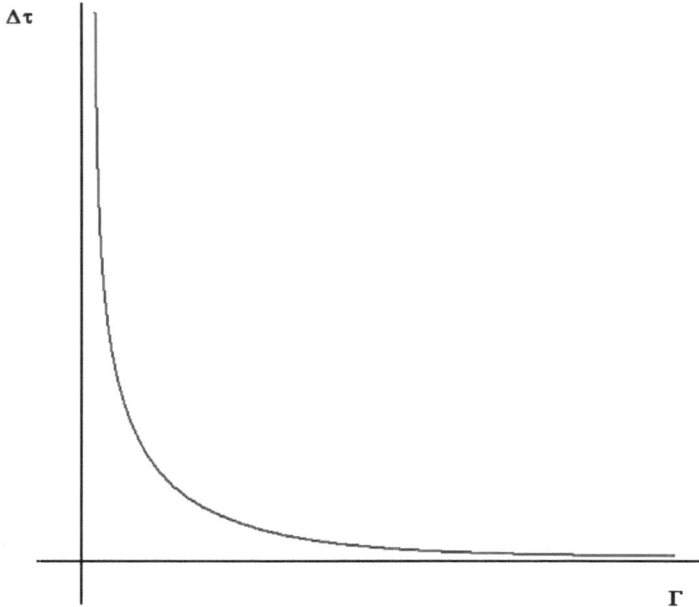

Γ

The rate of passage of time, t, is described here as the succession of quantized time intervals:

$\Delta\tau,\Delta\tau,\Delta\tau,\Delta\tau,\Delta\tau,\Delta\tau,\Delta\tau,\Delta\tau,\Delta\tau,\Delta\tau,\Delta\tau,\Delta\tau,\Delta\tau$ ...... *etc.*

so that at zero gravity or as $\Gamma \rightarrow$ *zero,* $\Delta\tau$ $\rightarrow$ *infinity*

                          and as $\Gamma \rightarrow$ *infinity,* $\Delta\tau \rightarrow$ *zero*

The contraction/expansion of quantized length $\Delta\lambda$ is also described here as a function of gravitational field strength Γ:

Graph of Quantised Length $\Delta\lambda$ versus Gravitational Field Strength Γ

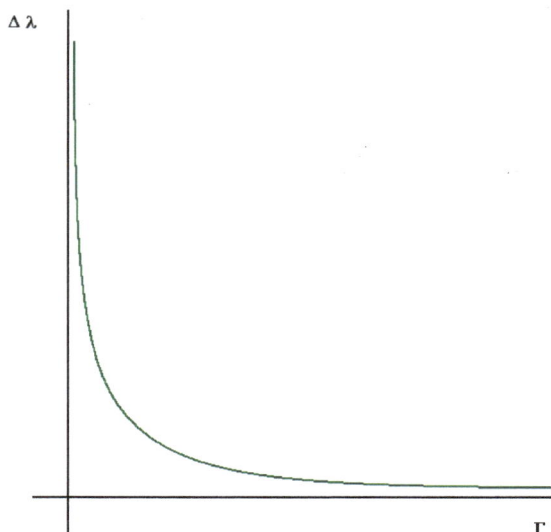

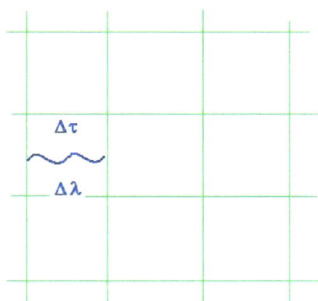

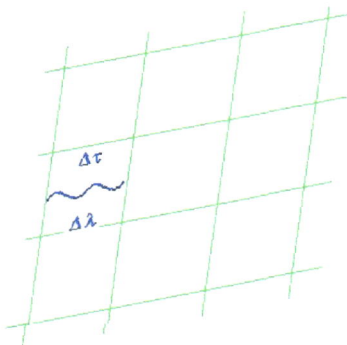

244

Interestingly, the ratio $\Delta\lambda/\Delta\tau$ remains fairly constant! A photon traverses a quantum of length in a quantum of time which is the value *c*, the speed of light we measure on the Earth and deem to be constant in "empty space" (a term which I associate as being meaningless!)

I said that photons are space creators and it is in this context that we see the expansion of space between galaxies or "space slip". Not only does space expand but it is there that photons interact to produce all the building blocks of matter i.e those sub-atomic particles that eventually combine to form protons, neutrons, electrons and atoms of the most fundamental element, hydrogen. (I hypothesise that it is here also that creation of the 'primal pulse' in pairs or, more likely, in triplets is constantly under way as a natural process.) All other elements as we know are forged or synthesised by nuclear reactions inside stars; eventually to be spat out into space in the death throws of the star which we call a super nova. Great clouds of gas then coalesce where carbon reacts with hydrogen oxygen and nitrogen to form organic molecules that we see as the building blocks of life. These clouds called nebulae (singular: nebula) are thousands of lightyears in dimension and last hundreds of millions of years before gradually condensing under gravity to form new star systems, sometimes with planets. The rest regarding the origin of life and its story of evolution are narrated later.

We have observed that the passage of photons across space undergo a curved trajectory as they pas by a massive object like a star. We attribute this curved path to the curvature of space due to high gravitational field:

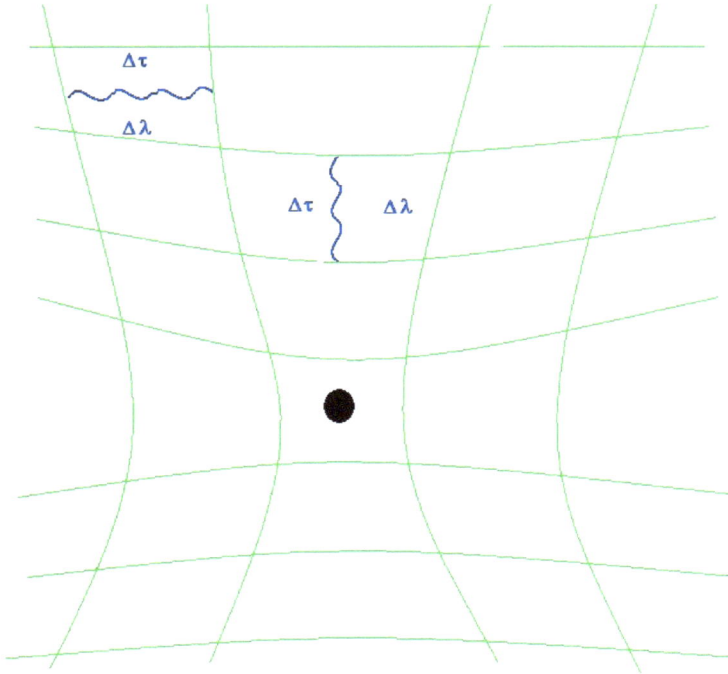

My interpretation here, following the line of discussion, is that both the quantised description of time and length are compressed with increasing gravitational field about the star. I said earlier that the velocity of light remains fairly constant from the relation $\Delta\lambda/\Delta\tau$. But I have no experimental evidence that it is constant, especially at the extreme where gravitational field approaches infinity: $\Gamma \rightarrow$ infinity. We know that the value of c is smaller the denser the medium it traverses i.e when light passes through air, water or clear glass. Also, photons that come too near a substantial black hole will be sucked in altogether by the extreme intensity of its gravity. So I suspect that *there may exist some departure from its constancy at the extremes of gravitational field strength.*

[Interestingly here, if we could artificially connect two points with a thin thread or bubble of "no space" $\Gamma_{\Lambda\Omega}$, the science fiction idea of instant traversal and arrival would be possible!]

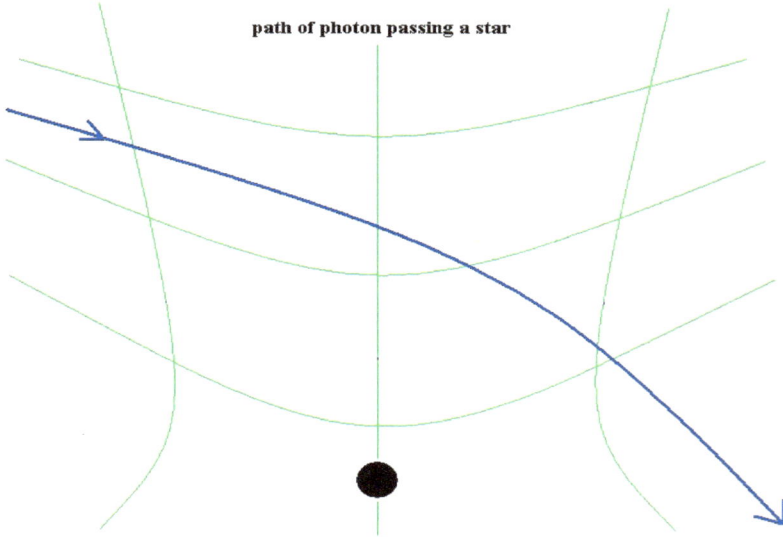

path of photon passing a star

It is this phenomenon of the gravitational refraction of light due to the curvature of space that has enabled astronomers to identify the existence and position of black holes in our galaxy where the background stars appear to be out of alignment. Observations are taken six months apart of a region of space as the Earth orbits the sun and comparisons of the background field of stars made.

To round off then, Fred Hoyle and George Gamow seem both to be correct if we describe a 'Universe La Grande' that ebbs and flows in a continuous everlasting cycle, remaining conducive to the fundamental law of constancy of total energy. But that first ripple I think will remain elusive to fully comprehend and we may never be able to confirm by an experiment the precise nature of the first move. Steady State and Big Bangs will remain theories inseparable from conjecture and deeper questions!

Appendix G   The Amino Acids, the Building Blocks for Life

Below are the twenty amino acids found in all living things. Remember earlier we said that this is a small sub-set of hundreds of the smaller amino acids already known and that many have been found and identified in meteorites. This would date some of these several billion years old indicating that they were formed before the Earth and Solar System. This leads us to speculate on the possibility that the first replication structure, a primitive DNA, was created elsewhere in the Universe, most likely within the nebulae. Both the age and enormous dimension of nebulae providing a soup of carbon compounds, increases the chance of a primal life-form being created there rather than on the planet. The Earth was likely seeded from this first 'Mycoplasma Nebulum' from which all living things thence evolved.

Neutral Amino Acids

Alanine	A	CH_3CHCO^- with $C=O$, NH_3^+

$$CH_3 \overset{O}{\underset{NH_3^+}{C}HCO^-}$$

Glycine	G

$$CH_2CO^- \;\; NH_3^+ \;\; O$$

Asparagine	N

$$H_2N\overset{O}{C}CH_2\overset{}{\underset{NH_3^+}{C}}HCO^-$$

Isoleucine	I

$$CH_3CH_2\underset{CH_3}{C}H\overset{O}{C}HCO^- \;\; NH_3^+$$

Cysteine	C

$$HSCH_2\overset{O}{\underset{NH_3^+}{C}}HCO^-$$

Leucine	L

$$CH_3\underset{CH_3}{C}HCH_2\overset{O}{C}HCO^- \;\; NH_3^+$$

Glutamine	Q

$$H_2N\overset{O}{C}CH_2CH_2\overset{O}{\underset{NH_3^+}{C}}HCO^-$$

Methionine	M

$$CH_3SCH_2CH_2\overset{O}{C}HCO^- \;\; NH_3^+$$

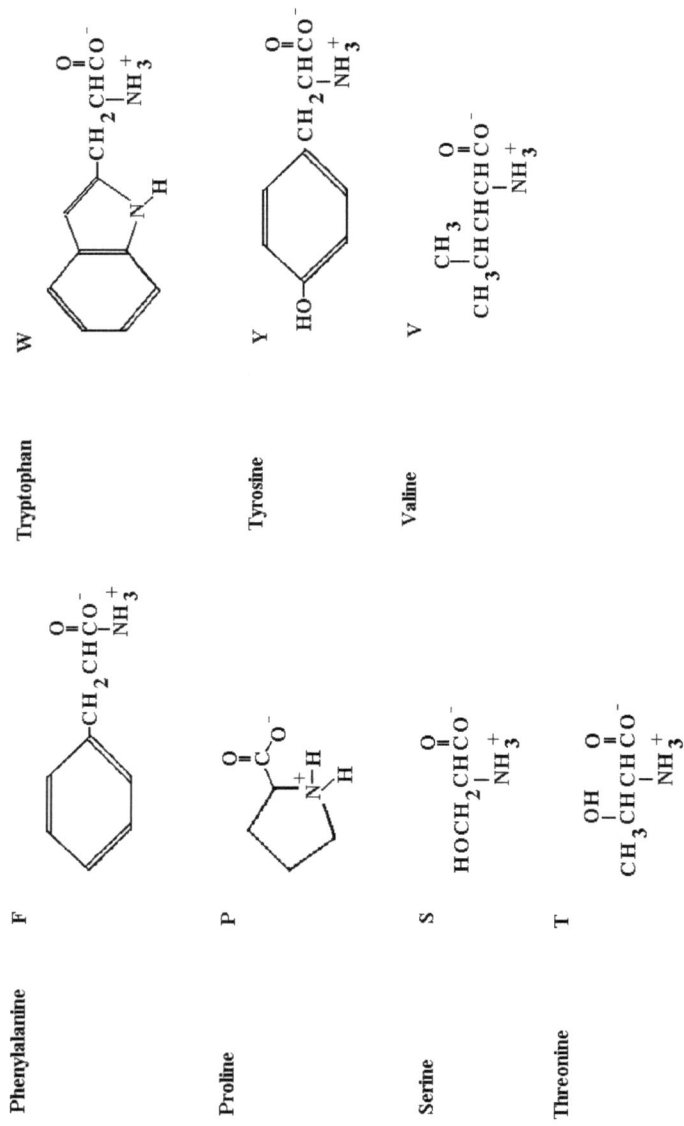

Phenylalanine F

Tryptophan W

Proline P

Tyrosine Y

Serine S

Valine V

Threonine T

Acidic Amino Acids

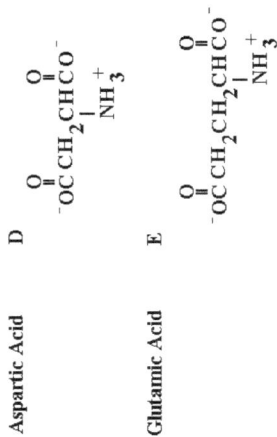

Aspartic Acid   D

$$\overset{O}{\underset{}{\|}}\ \ ^{-}OC\,CH_2\,\underset{\underset{NH_3^+}{|}}{CH}\overset{\overset{O}{\|}}{C}O^-$$

Glutamic Acid   E

$$^{-}OC\,CH_2\,CH_2\,\underset{\underset{NH_3^+}{|}}{CH}CO^-$$

Basic Amino Acids

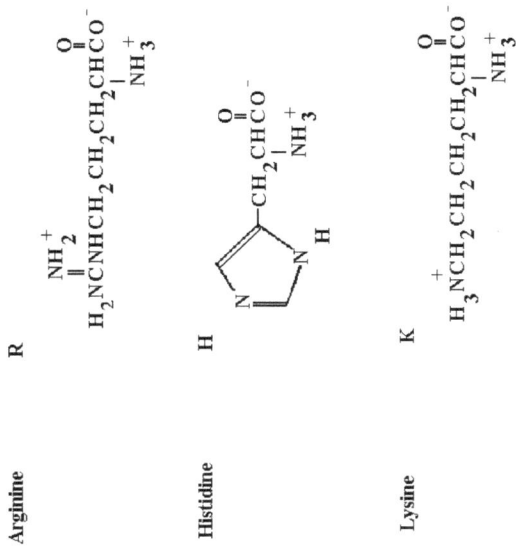

Arginine   R

$$H_2N\underset{\underset{NH_2^+}{\|}}{C}NHCH_2\,CH_2\,CH_2\,\underset{\underset{NH_3^+}{|}}{CH}CO^-$$

Histidine   H

$$CH_2\underset{\underset{NH_3^+}{|}}{CH}CO^-$$

Lysine   K

$$H_3NCH_2\,CH_2\,CH_2\,CH_2\,\underset{\underset{NH_3^+}{|}}{CH}CO^-$$

252

It has been estimated that approximately eighty five percent of all matter in the universe has not been detected, but this matter must be present for the universe to 'hang together' so to speak in terms of gravity. Without it, galaxies would fly apart at a far greater rate than observation shows us. Dark matter is postulated to explain such things as the flat rotation curves of spiral galaxies; gravitational lensing of background objects by galaxy clusters. We know there exists a small amount of baryonic dark matter. To overcome this problem cosmologists have invented the presence of 'dark matter' i.e matter that cannot to date be detected but is thought to exist in the far reaches of intergalactic space as well as in galaxies.

The vast majority of the dark matter in the universe is believed to be nonbaryonic, which means that it contains no atoms and that it does not interact with ordinary matter via electromagnetic forces. The nonbaryonic dark matter includes neutrinos, which were discovered to have mass in recent years, and may also include hypothetical entities such as axions, or supersymmetric particles. Unlike baryonic matter, nonbaryonic dark matter did not contribute to the formation of the elements in the early universe, at least not in a direct way. The density of ordinary baryons and radiation in the universe is estimated to be equivalent to about one hydrogen atom per cubic meter of space. Thus only about 4% of the total energy density in the universe, relying on gravitational effects, can be seen directly. About 22% is thought to be composed of dark matter. The remaining 74% is thought to consist of dark energy, an even stranger component, distributed diffusely in space.

I term a particle of dark matter as the mota. It is likely that these particles are considerably smaller than the proton, possibly of order one thousandth of the mass of the proton. They would have been created in the early universe from fundamental quanta or photons (see Appendix E, The First Move). However, instead of forming a proton, neutron or electron, the combination forms a tight bundle of elementary matter not able to co-join to form molecules or attract electrons into energy orbits. Because of the latter property, a mota cannot absorb photons or emit photons as do electrons around atoms. For this reason motas cannot be seen. So what happens to light passing through a cloud of these particles? It can refract but must come exceeding close to a mota particle; or it can bounce (reflect) from a direct collision.

It may be that the mota and vast clouds of this dark matter preceded atoms, stars and galaxies. Without motas it may not have been possible for protons, neutrons and electrons to have been formed. Further, the process of creation of hydrogen atoms is most likely still going on in intergalactic space with the mota playing a key role in their formation. Alternatively is that we need a modification of the theory of gravity to explain these effects!

Are there motas existing in rocks on Earth or entrapped as part of meteoritic material? Again most likely. It's just that no-one has looked for them. Seek and ye shall find!

And if we never find them? Well, it means that our whole understanding of gravity and space needs re-examining. A finite but unbounded universe in itself might exude the mathematical key to the conundrum of dark matter and dark energy!

If non-baryonic dark matter does exist, then our world and the people in it will be removed even farther from the centre of the Universe. Dr Sadoulet tells the New York Times, "It will be the ultimate Copernican revolution. Not only are we not at the centre of the universe as we know it, but we aren't even made up of the same stuff as most of the universe. We are just this small excess, an insignificant phenomenon, and the universe is something completely different"

Waiting for Daddy

Bibliography

A Compendium of Data on Global Change Carbon Dioxide Information Analysis Center. Oak Ridge National Laboratory,
U.S. Department of Energy. <http://cdiac.esd.ornl.gov/trends/trends.htm>

A Shift in Seasons,National Geographic, Sept. 2004, 44-45.

A Short History of Nearly Everything, Bryson Bill, Broadway Books,2003

A Short History of Chemistry, Asimov Isaac, Greenwood Press , 1982

Abraham Darby and the Coalbrookdale Furnace, Ironbridge Institute, University of Birmingham and the Ironbridge Gorge Museum Trust, UK

AlarmingAcceleration in CO2 Emissions Worldwide, Field CB,Science Week 2007 Carnegie Institute for Science

American Chemical Society

An Inconvenient Truth, Al Gore

An Obligately Photosynthetic Bacterial Anaerobe from a Deep Sea Hydrothermal Vent, Beatty J. Thomas,
Proceedings of the National Academy of Sciences (June 2007).

Antarctic Mass Rates from GRACE,Chen, J. L., C. R. Wilson, D. D. Blankenship, and B. D. Tapley, Geophysical Research Letters, 33:L11502,2006

Assessing the Potential for Geological Carbon Sequestration in the UK , Tyndall Centre

Baptistina- The Dinosaur Destroying Meteorite, Dunham Will, Reuters W, Sep 2007

Bermuda Biological Station for Research, St. George's, Bermuda

Big Bang: The Origin of the Universe, Singh S,Fourth Estate, 2004

Biogeochemical Evolution- Course Notes, Huskey Robert, University of Virginia 2004

Biogeochemical Evolution- Course Notes, Jürgen Schiever, University of Virginia 2002

British Energy, http://www.british-energy.com/

Calcium, Office of Dietary Supplements , NIH Clinical Center , National Institutes of Health

Canada Pumps CO2 Underground (Weyburn Oil Field), Seattle Times, 20 February 2004,

Carbon Dioxide Sink, Wikipedia

Characterisation of the Reconstructed 1918 Spanish Influenza Pandemic Virus,TuMpey T M et al, Science 10, 2005, 77-80

Chemical and Engineering News

Chemistry and geochemistry of the ocean

Chemistry Between the Stars, New Scientist 79, 400-403 (1978).

Climate Change, Human Impacts, and the Resilience of Coral Reefs," Science, 301:929-933

Climate Change: Managing Forests After Kyoto, Schulze, E.-D., C. Wirth, and M. Heimann.289: 2058-2059   2000.

Climate Change: Risk, Ethics and the Stern Review, Stern N, Science 317, 2007, 203-4

Cosmic Hide and Seek: the Search for the Missing Mass,Miller,Christopher M. 1995. http://www.eclipse.net/~cmmiller/

Cosmology, Abell, George O., and Marc Davis. McGraw-Hill Encyclopedia of Science and Technology. 7th ed. New York: McGraw-Hill, 1992.

Dal Big Bang ai buchi neri,Hawking S. ,Rizzoli, Milano 1992

Deep Simplicity: Chaos Complexity and the Emergence of Life , Gribbin John, Penguin Press Science

Department of Energy, www.netl.doe.gov/publications/carbon_seq/refshelf.html

Encyclopedia Brittanica

Evolution of the Biosphere- Course Notes, Huskey Robert, University of Virginia 2003

Evolution; Triumph of an Idea, Zimmer Carl, Harper Perrenial, 2001

Evolutionary Timeline, Brant Neil, 2005

Following the Trail of Light, Calvin Melvin, American Chemical Society 1992

Gas Solubility in Sea Water, Yuji Sano and Naoto Takahata, Center for Advanced Marine Research, Ocean Research Institute,
The University of Tokyo, Tokyo 164-8639

General Chemistry, Pauling, Linus, 1970 ed. Dover Publications

GISS Surface Temperature Analysis,Hansen, J., et al., 2006, Goddard Institute for Space Studies, on line [http://data.giss.nasa.gov/gistemp/]

Glaciers and the Changing Earth System: a 2004 Snapshot,Dyurgerov, M. B., and M. F. Meier, 2005, INSTAAR, on line

Global Warming and the Stability of the West Antarctic Ice Sheet, Oppenheimer, M., 1998, Nature, 393:325-332.

Global Warming, Johnston Wm. Robert

Gravity Lenses: A Focus on the Cosmic Twins, Falco, Emilio and Nathaniel Cohen. Astronomy. July 1981: 18-22.

Greenpeace International, www.greenpeace.org

History & Manufacture of Portland Cement: Joseph Aspdin to the present, Fact sheet

Holocene deglaciation of Marie Byrd Land, West Antarctica, Stone, J. O., et al., 2003, Science, 299:99-102.

Home Alone, Law Tom, Longership Publishing Australia 2008

Ice-sheet and Sea-level Changes, Alley, R. B., P. U. Clark, P. Huybrechts, and I. Joughin,Science, 310:456-460, 2005

Image (AS17-148-22727) courtesy of the Image Science & Analysis Laboratory, NASA Johnson Space Center, http://eol.jsc.nasa.gov.

Important! Why Carbon Sequestration Won't Save Us, Michael Graham Richard, Gatineau, Canada 07.31.2006

Inventory of U.S. Greenhouse Gas Emissions and Sinks: 1990-2002, EPA (2004)

Iridium Spikes and the Dusk of the Dinosaurs, Alvarez Walter

Irregular Gippsland Peace Letter, Gardner P D, Ngarak Press, Australia

John Daltons Notation, John H. Lienhard, University of Houston

Johnston Dr Wm Robert, www.johnstonarchive.net

Life: The Science of Biology, 4th Edition Purves et al., by Sinauer Associates (www.sinauer.com) and WH Freeman (www.whfreeman.com).

National Geographic News, John Pickrell, July 15, 2004
National Oceanic and Atmospheric Administration
Nature Magazine
New Scientist
New York Times Magazine
Notebook From Vietnam, McPhee Bruce, Personal Political Writings, 2003-2008
Nuclear Decommissioning Authority, http://www.nda.gov.uk/
Nuclear Islam and Other Stories, Law Tom, Longership Publishing Australia 2008
Organic Chemistry, McMurry John ,Thomson Brooks/Cole, 2004
Organic Molecules in the Interstellar Medium, Comets and Meteorites, Ehrenfreund Pascale and Charnley S.B.,Leiden Observatory,
Review of Astronomy and Astrophysics Vol 38 2000 pp 427-483
Origin of Life: the RNA World, Gilbert W319, Nature 319, 1986 p618
Origin of Limestone Caves, Steven A. Austin
Plows, Plagues and Petroleum; how Humans Took Control of Climate, Ruddiman W F, Princeton University Press, 2005
Poems by Chas Rose 1975- 1995 Longership Publishing Australia 2009
Polycyclic Aromatic Hydrocarbons, Harvey R.G., Wiley-VCH New York 1997.
Pseudoscience Going up in Smoke, Washington Times article, 1998
PseudoScience- Lower Stephen (retired), Notes from Dept of Chemistry, Simon Fraser University Burnaby Vancouver
Reduction in Himalayan Snow Accumulation and Weakening of the Trade Winds over the Pacific since the 1840s," Zhao, H., and G. W. K. Moore,
2006, Geophysical Research Letters, 33:L17709
Reflections on the Chemico-biological Role of Interstellar Black Material and the Appearance of Life in the Universe,
Ehrenfreund Pascale and Charnley S.B. [www.tightrope.it/nicolaus/index.htm]
Revelation 6: The 7 Seals; Holy Bible, King James Version, London 1701
Science Magazine
Scientific America
Scotland Left Off New Nuclear Map, Barnes Eddie and Brady Brian, Scotland on Sunday, 20 May 2007
Selected Food Sources of Calcium , Office of Dietary Supplements , NIH Clinical Center , National Institutes of Health
Skeptical Inquirer- The Magazine for Science and Reason, published by CSICOP.
Sleipner CO2 Injection Project
State of the Cryosphere: Is the Cryosphere Sending Signals About Climate Change?, National Snow and Ice Data Center, 14 March 2005, NSIDC,
[http://nsidc.org/sotc/glacier_balance.html]
Taking the High (Fuel Economy) Road, World Resources Institute, 2004, [http://pdf.wri.org/china_the_high_road.pdf].
The 11th Hour, DVD
The Carbon Age,Roston Eric , Walker and Company, 2008
The Carbon Cycle and Atmospheric Carbon Dioxide. In Climate Change: The Scientific BasisPrentice, Colin I et al.
—Contribution of Working Group I to the Third Assessment Report of the Intergovernmental Panel on Climate Change,
eds. John T. Houghton et. al., Cambridge, U.K.: Cambridge University Press, (2001):183–237.
The Chemical Evolution of the Atmosphere and the Ocean, Holland H.D., Princeton Univ. Press, Princeton 1984.
The Citric Acid Cycle, Krebs Hans, Nobel Lectures Physiology or Medicine 1942-1962, Amsterdam, Elsevier Publishing Company 1964
The Depletion of Calcium in Forest Ecosystems of the Northeastern USA, Hubbard Brook Experimental Forest Research Group, 2005
The Emergence of Cells During the Origin of life, Chen irene, Science 314 2006, 1558-1559
The Emergence of Everything, Morowitz H, OUP 2002
The Emergence of Life: From Chemical Origins to Synthetic Biology, Luisi Pier Jul 2006
The Energetics of Carbon Dioxide Capture in Power Plants, Dr. Gerold Göttlicher ,
The National Energy Technology Laboratory (NETL) of the U.S. Department of Energy (DOE)
The First Organisms, Cairns-Smith A. G, Scientific American 253, 1985 90-100
The Life Cycles of Stars, Chandrasekhar
The Loss of Ice Shelves from the Antarctic Peninsula, May 2000, British Antarctic Survey,
[http://www.antarctica.ac.uk/Key_Topics/IceSheet_SeaLevel/ice_shelf_loss.html]
The Omega Point: The Search for the Missing Mass and the Ultimate Fate of the Universe, Gribben, John. . New York: Bantam, 1988.
The Origin of Atmospheric Oxygen, - Course Notes, Huskey Robert, University of Virginia 2003
The Origin of the Species and Voyage of HMS Beagle, Darwin Charles, Penguin Reprint
The Periodic Table, Levi Primo,Schoken Books, 1984
The Population Bomb, Ehrlich, P., 1968,
The Potential for Sea Level Rise: New estimates from Glacier and Ice Cap Area and Volume Distributions, Raper, S. C. B., and Braithwaite R. J.,
8 March 2005, Geophysical Research Letters, 32:L05502.
The Rise of Plants, Berner Robert,Science 276 1997 544-546
The Role of Markets in Forest-based Climate Mitigation, Clean Development Mechanism
The Shape of Carbon Compounds, Herz Werner, WA Benjamin 1963
The SmithsonianNASA Astrophysics Data System
The Universe, Past and Present Reflections, Hoyle Fred,Annual Review of Astronomy and Astrophysics 20, 1982, 1-36
The Weather Makers, Tim Flannery Tim.
Tracers in the Sea, Broecker, Wallace S., and Tsung-Hung Peng. Palisades, NY: Eldigio Press, 1982
Trends in Atmospheric Carbon Dioxide," Tans, P., 2006, NOAA ESRL, [http://www.cmdl.noaa.gov/ccgg/trends/].
U.S. Forest Facts and Historical Trends, United States Department of Agriculture - Forest Service. (2004).
UK Atomic Energy Authority, http://www.ukaea.org.uk/
US Coal Board
US Department of Energy
Water Encyclopedia
When will the Present Interglacial End?,"Kukla, G. J., and R. K. Matthews. 1972, Science, 178:190-191
World Coal Demand and Supply Prospects, Coal Industry Advisory Board Meeting with IEA Governing Board December 2003
World Nuclear Association, http://www.world-nuclear.org/
World Primary Energy Production Trends, http://www.eia.doe.gov/iea/overview.html

Other Books by Tom Law:

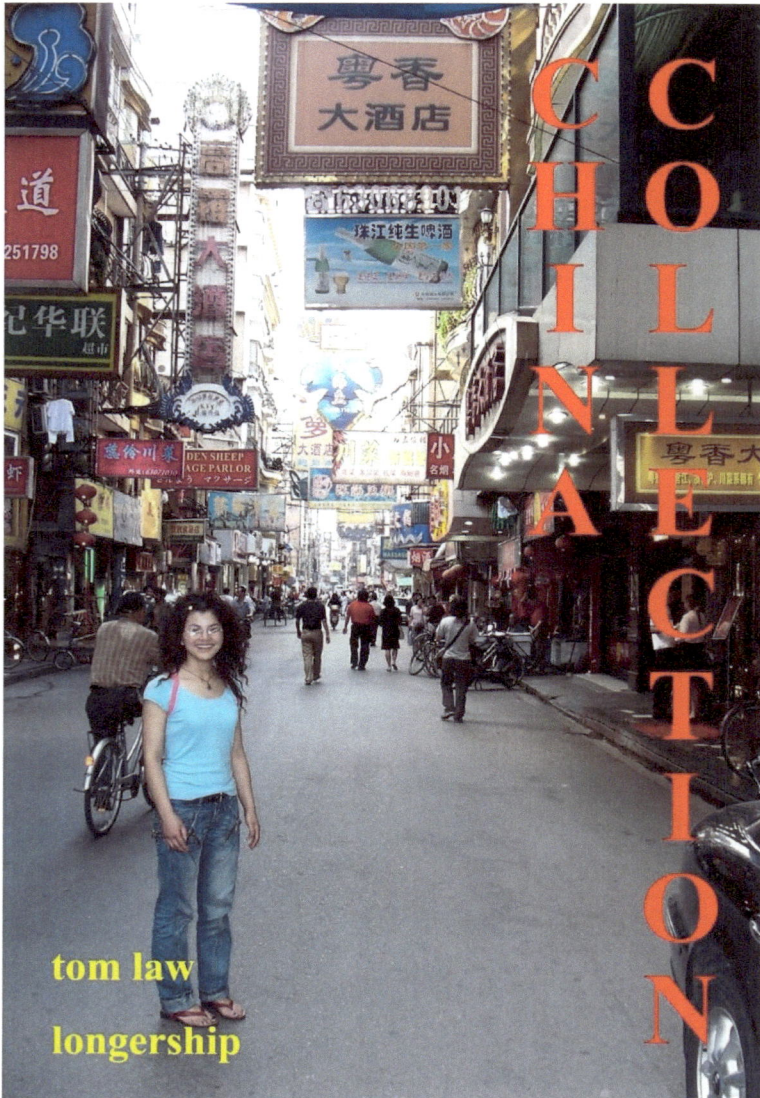

CHINA COLLECTION

tom law

longership

boy in blue raincoat

tom law   LONGERSHIP

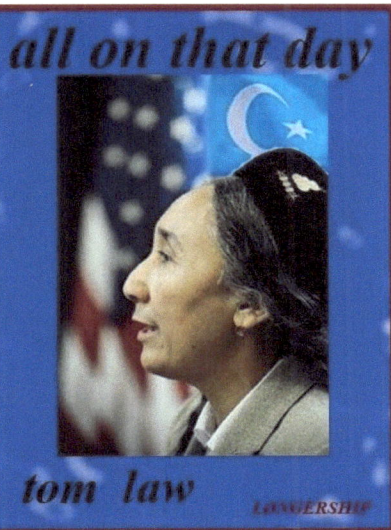

all on that day

tom   law   LONGERSHIP

S C O T L A N D S

C H O I C E

260

From Russia

with Love

Tom Law

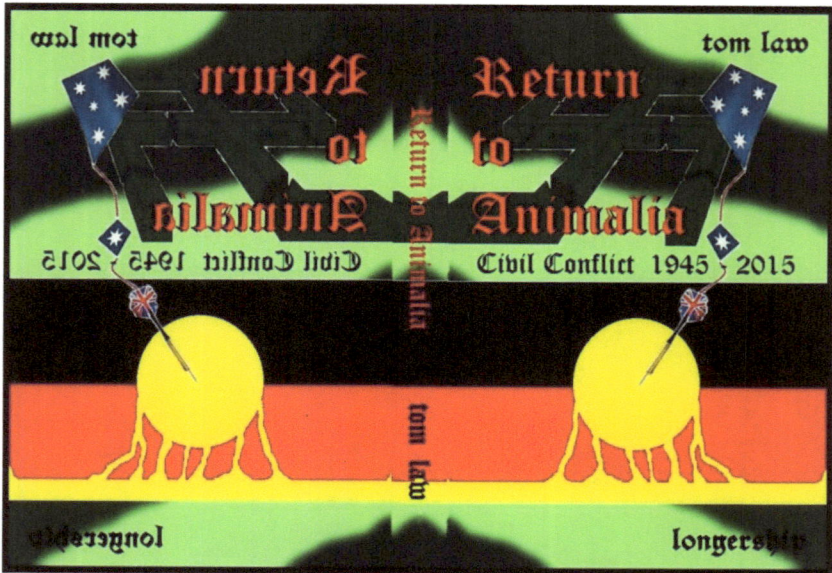

Return to Animalia

Civil Conflict 1945 - 2015

tom law

longership

www.ingramcontent.com/pod-product-compliance
Lightning Source LLC
Chambersburg PA
CBHW040124270326
41926CB00001B/3